# Planning and Standard Operating Procedures for the Use of Rotenone in Fish Management

—

# Rotenone SOP Manual

This document was made possible by funds provided by the

U.S. Fish and Wildlife Service
Division of Fish and Wildlife Management
and Habitat Restoration

# Planning and Standard Operating Procedures for the Use of Rotenone in Fish Management

—

# Rotenone SOP Manual*


American Fisheries Society
Fish Management Chemicals Subcommittee

Authors

Brian Finlayson, California Department of Fish and Game†
Rosalie Schnick, Michigan State University
Don Skaar, Montana Fish, Wildlife and Parks
Jon Anderson, Washington Department of Fish and Wildlife
Leo Demong, New York State Department of Environmental Conservation
Dan Duffield, Forest Service, U.S. Department of Agriculture
William Horton, Idaho Department of Fish and Game
Jarle Steinkjer, Norway Directorate for Nature Management


*The applicator is responsible for conforming to the product labeling. This manual provides guidance on the safe and effective application of rotenone and is intended for the use of fish biologists and fishery managers in the United States and Canada. Related rotenone information is available at www.fisheries.org/units/rotenone. The manual will be updated as the situation warrants; check for current version at this website.


†Author to whom correspondence should be submitted:
Brian Finlayson
2271 Los Trampas
Camino, CA 95709
briankarefinlayson@att.net



**American Fisheries Society**
**5410 Grosvenor Lane**
**Bethesda, Maryland 29814-2199**
**USA**

A suggested citation format for this book follows.

Finlayson, B., R. Schnick, D. Skaar, J. Anderson, L. Demong, D. Duffield, W. Horton, and J. Steinkjer. 2010. Planning and standard operating procedures for the use of rotenone in fish management—rotenone SOP manual. American Fisheries Society, Bethesda, Maryland.



Printed in the United States of America on acid-free paper.

Library of Congress Control Number 2010905480
ISBN 978-1-934874-17-2

American Fisheries Society Web site address: *www.fisheries.org*

American Fisheries Society
5410 Grosvenor Lane, Suite 100
Bethesda, Maryland 20814
USA

This manual is dedicated to Richard L. Cailteux (1964–2008).

In memory of our colleague Rich for his friendship and insight on piscicides and his organizational and editorial skills during the writing of numerous publications of the Fish Management Chemicals Subcommittee.

# Table of Contents

# ACKNOWLEDGMENTS

The *Rotenone SOP Manual* was made possible by funds provided in January 2007 by the U.S. Fish and Wildlife Service, Division of Fish and Wildlife Management and Habitat Restoration (Grant 98210-4-G735), and requested by the American Fisheries Society's Fish Management Chemicals Subcommittee. The authors would like to thank the following for their comments and suggestions on the manual: Robert Adair (U.S. Fish and Wildlife Service), Pat Clancey (Montana Fish, Wildlife & Parks), Stephen Covell (U.S. Forest Service), Dirk Helder and Tom Steeger (U.S. Environmental Protection Agency), Terry Hubert (U.S. Geological Survey), Stephen Hurst (New York State Department of Environmental Conservation), Paul Janssen (Idaho Department of Fish and Game), Aaron Lerner (American Fisheries Society), Don Wiley (Utah Division of Wildlife Resources), and Kirk Young (Arizona Game and Fish Department). The authors also acknowledge the continued support of Hannibal Bolton (U.S. Fish and Wildlife Service) and Larry Riley (Arizona Game and Fish Department) in the production of this manual.

# PREFACE

Rotenone was first used as a piscicide in the United States and Canada in the 1930s. Prior to the passage of the amendment to the Federal Insecticide, Rodenticide and Fungicide Act of 1970, rotenone and other pesticides were regulated in the United States by the U. S. Department of Agriculture, and rotenone was first registered in 1947. In 1970, pesticides became regulated by the U.S. Environmental Protection Agency. Registration now requires physical, chemical, public health and environmental data to demonstrate effectiveness at appropriate concentrations of the intended use and to allow for assessments of risk-estimating the impacts to human health and the environment. In 1988, all pesticides, including rotenone, registered before November 1, 1984 were put into a reregistration process requiring the generation of data to support continued registration. After the necessary risk assessments for rotenone were completed, the U.S. Environmental Protection Agency issued the Reregistration Eligibility Decision in March 2007 (EPA 738-R-07-005). As a condition of the reregistration requirements, a manual was requested by EPA that contained procedures, specifically on how to minimize nontarget exposure and effects and to provide guidance on the new, reasonably complex label use conditions. With the approved reregistration came several significant technical changes in how rotenone will be used as a tool in fish management. These changes are incorporated into Standard Operating Procedures (SOPs) in this manual that provide guidance on how to comply with the label, and use rotenone in a safe and effective manner. Many of the SOPs in the manual (SOPs 5, 6, 7, 8, 9, 10, 13 and 16) are referenced on the label and must be followed to the extent dictated by the wording. The manual contains other SOPs which, even though considered largely advisory, complement the safe and effective use of rotenone and should be understood and followed. The manual also provides guidance on project planning procedures needed for rotenone use. The American Fisheries Society's Fish Management Chemicals Subcommittee, in cooperation with the Rotenone Task Force (coalition of rotenone registrants) and U.S. Environmental Protection Agency developed the manual.

The manual contains four sections beginning with an introduction that orients the reader to a brief history of rotenone registration, new label directions, use as a fishery management tool, information on formulations and environmental fate, public health and concerns, and an overview of the manual structure. The second section provides general guidance on rotenone project planning procedures and how the Standard Operating Procedures are used in successful planning. The third section contains Standard Operating Procedures that limit rotenone exposure and effects and compliment the label. The fourth and final section provides a complete list of references. Detailed instruction on rotenone application techniques, hands-on experience with application equipment, matching rotenone formulations to specific waters and examples of successful planning are provided in a week-long course given by the Fish Management Chemicals Subcommittee. Information on current piscicide course scheduling can be found at the American Fisheries Society's website www.fisheries.org/units/rotenone/.

# Common Abbreviations and Glossary of Terms

Absorption = movement of treated water or product through the skin or eyes

Active Rotenone = concentration of active ingredient rotenone in surface water or other medium expressed as ppb (μg/L) or ppm (mg/L)

AF = surface acre of water one foot deep

AFS = American Fisheries Society

a.i. = percentage (%) of active ingredient rotenone in liquid or powdered commercial formulation

atm = atmosphere; used in Henry's Law constant as a measure of the solubility of a gas in liquid

BCF = bioconcentration factor that is the potential for a substance to accumulate in living biological tissue and is normally expressed as the ratio of rotenone concentration in tissue to water

CAS = Chemical Abstracts Service number for a unique chemical

CFR = Code of Federal Regulations

CHCP = Comprehensive Hazard Communication Plan required by OSHA's HCS.

Control = reduction of fish populations or fish species

CWA = Clean Water Act

Deactivation = the processes of hydrolysis, photolysis, and chemically induced oxidation that degrade rotenone to nontoxic concentrations. This process has been previously referred to as detoxification, neutralization and degradation.

Dispersant = a substance that assists in spreading another substance

EA = environmental analysis

Emulsifier = generally a petroleum-based substance in water; a substance used to stabilize the suspension of one liquid in another

EPA = U.S. Environmental Protection Agency

Eradication = elimination of whole fish populations or fish species from distinct habitats or bodies of water

ESA = Endangered Species Act

F = female animal

FFDCA = Federal Food, Drug and Cosmetic Act

FIFRA = Federal Insecticide, Fungicide, and Rodenticide Act

FMCS = Fish Management Chemicals Subcommittee

FMP = Fish Management Plan

$ft^3$ = cubic feet

FS = Forest Service of the U.S. Department of Agriculture

FWS = Fish and Wildlife Service of the U.S. Department of the Interior

GPM = gallons per minute

HCS = Hazard Communication Standard required by OSHA provides employees with the identities and hazards of chemicals to which they are exposed

HP = horse power

Hydrolysis = the degradation of a substance through reaction with water

ICS = Incident Command System

Ingestion = swallowing treated water or product

Inhalation = breathing volatile product vapors and dust

$K_d$ = sediment partition coefficient

$K_{ow}$ = octanol-water partition coefficient

$LD_{50}$ or $LC_{50}$ = a statistically derived estimate of a dose or concentration of a substance that would cause 50% mortality to the test population under specified conditions

LOAEL = lowest observable adverse effect level from a toxicity study; also referred to as LOEL

$m^3$ = cubic meters

M = male animal

MATC = maximum acceptable toxicant concentration = $(\text{NOAEL} \bullet \text{LOAEL})^{1/2}$

MED = minimum effective dose

mg/kg = milligrams per kilogram = ppm

mg/L = milligrams per liter = ppm

mol = molar

MSDS = Material Safety Data Sheet required by OSHA and used as part of the CHCP

NEPA = National Environmental Policy Act

NIOSH = National Institute of Occupational Safety and Health rating of safety equipment

NOAA = National Oceanic and Atmospheric Administration of the U.S. Department of Commerce

NOAEL = no observable adverse effect level from a toxicity study; also referred to as NOEL

NPDES = National Pollutant Discharge Elimination System

OSHA = Occupational Safety and Health Administration or Act

Oxidation = the degradation of a substance by uniting with oxygen

Pa = pascal unit of force equal to one Newton per square meter ($m^2$)

Pesticide = any substance or mixture of substances intended for preventing, destroying, repelling, or mitigating any pest and registered under the authority of FIFRA

Photolysis = the degradation of a substance caused by exposure to light

PIS = primary irritation score in ocular toxicity studies

Piscicide = chemical toxic to fish that is used to control, eradicate, or sample fish populations and registered as a pesticide under the authority of FIFRA

PSI = pounds per square inch rating for pressure

ppb = parts per billion; equivalent to µg/L (micrograms per liter) or µg/kg (micrograms per kilogram)

PPE = personal protective equipment required by the label and MSDS to mitigate for the four possible routes of pesticide exposure; absorption (skin and eyes), inhalation, ingestion

ppm = parts per million; equivalent to mg/L (milligrams per liter) or mg/kg (milligrams per kilogram)

RED = Reregistration Eligibility Decision

Restricted Use Pesticide = a pesticide restricted for a specific reason usually related to safety; to be used only under the direct supervision of a Certified Applicator

RPM = revolutions per minute

Service Containers = a container, other than original product container, that is approved for rotenone storage and contains (1) name and address of person or firm responsible for container, (2) identity of the pesticide in the container and (3) signal word from the original container

Signal Word = a rating of the acute health hazard of a pesticide on its label that ranges from “Danger” (Toxicity Category I or extremely harmful) to “Warning” (Toxicity Category II or moderately harmful) to “Caution” (Toxicity Category III)

SOP = Standard Operating Procedure

$t_{1/2}$ = half-life or the time period in which half of an amount of substance degrades

Tolerances = residue concentrations of a chemical that are permitted by regulatory agencies in water or food consumed by humans

Toxicity Category = see Signal Word

μg/kg = micrograms per kilogram = ppb

μg/L = micrograms per liter = ppb

Undesirable fish = species of fish designated by fisheries managers as undesirable in certain bodies of waters

USDA = U. S. Department of Agriculture

# 1 Introduction

The *Rotenone SOP Manual* is designed to provide fishery managers and others with procedures needed for carrying out restoration projects with rotenone in an effective and safe manner while meeting existing laws and regulations of all regulatory jurisdictions. The authors and reviewers of the manual represent a wide range of knowledge and experiences in using rotenone from four federal and seven state natural resource agencies throughout the United States and from Norway.

## 1.1 Registration, Label Directions and SOPs

Over the past several years, the use of rotenone has become a concern to a variety of interests including environmental and animal rights groups (Williams 2004; Finlayson et al. 2005). As a result, its use has been challenged, halted, or discouraged. In 1993, the AFS FMSC recognized a need to respond to increased concerns and established the "Rotenone Stewardship Program." In 2000, FMCS used FWS Division of Federal Aid administrative funds to prepare and produce a manual on rotenone use for fisheries managers entitled, *Rotenone Use in Fisheries Management—Administrative and Technical Guidelines Manual* (Finlayson et al. 2000). The *Rotenone SOP Manual* is a revision and expansion of the 2000 manual because EPA required it as part of the reregistration process for rotenone.

AFS FMCS became involved in the reregistration process for rotenone by reviewing individual Reregistration Eligibility Decision (RED) chapters on public health, environmental fate and effects, and occupational safety beginning in December 2005. AFS FMCS provided written and verbal comments and met with the registrants and EPA from 2005 throughout 2007 in resolving numerous issues and concerns over the reregistration of rotenone. The final RED was issued on March 31, 2007 and modified in a Response to Comments document on March 23, 2009 by the EPA (available at www.regulations.gov as EPA-HQ-OPP-2005-494-0036). Most significant for fisheries uses were the risk mitigation measures for rotenone listed in Table 14 (EPA 2007, p. 28–29) and corresponding label changes for end-use products listed in Table 16 (EPA 2007, p. 36–44). AFS FMCS met with the registrants and EPA in 2007 and 2008 to provide alternative label wording that provided feasible and effective risk mitigation. AFS FMCS believes that the current changes to the label and this *Rotenone SOP Manual* will result in the safe and effective use of rotenone for fish management while reducing the risk to and providing adequate protection of public health and the environment. Many of the changes to use directions involve reading and understanding the label. Following the procedures in this SOP manual will assist the applicator in understanding and implementing these important changes. Rotenone is classified as a Restricted Use Pesticide and as such, the Certified Applicator is responsible for insuring that all applicable laws and regulations and label instructions are followed.

A number of significant changes have been made to the use of rotenone. In summary, the changes (with corresponding SOPs if applicable) are:

- **Project Supervision and Safety**—The PPE requirements for powder and liquid formulations are different. Powder formulations require the use of a NIOSH-approved tight-fitting full-face cartridge or canister respirator, an approved helmet or hood-style respirator, or filtering face piece or half-face negative pressure respirator. Liquid formulations only require the use of a dust/mist respirator. Certified Applicators are required to remain on site until the treatment is completed and should receive training and have certain qualifications (SOPs 2 and 3).

- **Maximum Rotenone Treatment Levels**—The maximum treatment level in standing waters was reduced from 250 to 200 ppb rotenone (a.i.), and the maximum treatment level in flowing water was increased to 200 ppb rotenone (a.i.). For all applications, the selected treatment rate is based on response of target fish (or surrogate species) in a bioassay with site water (or in similar water) within the maximum level on the label (SOP 5).

- **Re-entry Requirements**—For treatment levels that are ≥ 90 ppb rotenone (a.i.), no reentry without PPE is allowed until levels decline to < 90 ppb rotenone (a.i.) (SOP 1). The length of time the area remains placarded is dependent on the treatment levels.

- **Monitoring Requirements for Treating Drinking Waters**—For treatment levels that are ≥ 40 ppb rotenone (a.i.), the Certified Applicator must inform drinking water users 7 to 14 days prior to treatment against the consumption of treated water (SOP 16).

- **Application of Rotenone**—All powder (with the exception of the powder/gelatin/sand mixture in SOP 13) formulations require the use of a semi-closed system for application (SOP 9). Similar application equipment is required for applying undiluted liquid rotenone from original product containers (SOP 8). The systems employ probes that are inserted into a bung hole of the drum (liquid) or plastic liner (powder) with a snug fit and the undiluted product is removed using suction created by a pump. Rotenone is generally applied below the surface, but liquid formulations may be applied to the surface of the water with backpack sprayers, drip cans, or hand-held or hand-directed spray nozzles (SOPs 11 and 12).

- **Transferring Liquid Rotenone Concentrate**—Transferring liquid rotenone from original product containers to service containers or other application devices is made in a plastic-lined, bermed area or other secondary containment area capable of recovering any spilled product as described (SOP 10).

- **Rotenone Use in Marine/Estuarine Environments**—The use of rotenone in marine/estuarine environments is not allowed.

- **Chemical Deactivation of Treated Flowing Water**—The flow of a stream or outflow of a treated lake beyond the treatment area now requires chemical deactivtion with potassium permanganate unless it is demonstrated to be unnecessary (SOP 7). The deactivation zone and other areas affected by the treatment are included in the definition of a project area (SOP 6).

- ***Rotenone SOP Manual***—This *Rotenone SOP Manual* was mandated by EPA and is provided for guidance on the safe and effective use of rotenone and to assist in following the label. The manual is considered labeling and should be consulted prior to treatment.

Only when rotenone is used according to the label is there a presumption that no unreasonable effects are expected to occur to humans or the environment. Thus, using rotenone in a manner inconsistent with its label is a violation of Federal and state laws. Violations may result in criminal penalties (up to a $5,000 fine and 6 months in jail per violation) when there has been an intentional or gross negligent misuse, or civil penalties (up to $25,000 per violation) or administrative remedies (revocation of license) when there has been an unintentional misuse. This manual provides guidance, but when in doubt, Certified Applicators should consult with their state or local pesticide control agency.

## 1.2 Fishery Management Tool

Rotenone continues to be a valuable tool in fisheries, without which many management options will be lost (McClay 2000; 2005; Finlayson et al. 2000). Fisheries managers rely on a wide variety of tools for the management and assessment of fish populations to maintain diverse and productive aquatic ecosystems and quality recreational fisheries. Piscicide application is the only method other than dewatering that can consistently eradicate undesirable fish communities or sample a portion of the entire fish population, including all species. Several species of threatened or endangered fish, most notably subspecies of the cutthroat *Oncorhynchus clarkii* and golden trout *O. mykiss aquabonita* and pupfish *Cyprinodon radiosus* owe their continued existence in part to rotenone. Rotenone has also made possible the eradication of several invasive species including white bass *Morone chrysops*, northern pike *Esox lucius*, snakehead *Channa argus* and tui chub *Gila bicolor* from nonnative habitats.

Rotenone has been used for centuries to capture fish in areas where rotenone-containing plants are naturally found.The piscicide was applied first to ponds and lakes, and then to streams by the early 1960s for either complete or partial reclamation (Schnick 1974). Rotenone was initially used in various powdered forms until emulsifiable formulations were developed that acted faster and were easier to handle and dispense. By 1949, 34 states and several Canadian provinces were using rotenone routinely for the management of fish populations (Solman 1950; Lennon et al. 1970). Rotenone is also used to sample fish populations for population assessments. In the past, rotenone was used as a natural insecticide on crops and livestock, but those uses have been voluntarily canceled by the registrants. In the past, humans took rotenone orally to control intestinal worms (Haley 1978).

As many as 30 piscicides have been used in fisheries management in the United States and Canada since the 1930s, but only four are currently registered for general or selective fish control or sampling. These products include the general piscicides, antimycin and rotenone, and the lampricides, Lamprecid® and Bayluscide®. Rotenone is the most extensively used piscicide in the United States (Cumming 1975; McClay 2000; 2005; Finlayson et al. 2002).

Certain fish are considered undesirable and may need to be removed because they impact desired fish through (a) competition, (b) predation, (c) genetic introgression, (d) harboring disease organisms, or (e) altering habitats (Krueger and May 1991; Ross 1991). Competition with nonnative salmonids has been implicated in the decline of many native inland salmonid species (Allendorf 1991; Krueger and May 1991; Finlayson et al. 2005). Predation on native fish has also been provided as a reason for removing fish, including the northern pikeminnow *Ptychocheilus oregonenis* (Zimmerman and Ward 1999), green sunfish *Lepomis cyanellus* (Lemly 1985), lake trout *Salvelinus namaycush* (Kaeding et al. 1996), and northern pike *Esox lucius* (California Department of Fish and Game 1991; 1997; 2007). Native fish stocks may be exposed to diseases that were not historically present in habitats occupied by native stocks (Krueger and May 1991). Baltic Sea stocks of Atlantic salmon *Salmo salar* are immune to the monogenean trematode parasite *Gyrodactylus salaris*, but it is fatal to indigenous Northern Sea stocks in Norway (Johnsen et al. 2008). Direct genetic effects include crosses between species that result in sterile hybrids and crosses that lead to introgression (Krueger and May 1991; Finlayson et al. 2005).

Fisheries managers may decide to use rotenone when fish communities have been disrupted by human activities (e.g., physical manipulations of natural waters, effects of pollution on natural production of fish species, demand for recreational fisheries, and introduction of exotic species into surface waters). The primary reasons for piscicide use have changed over the years. Originally, piscicides were mainly used to control undesirable fish populations so that sport fish could be stocked and managed for recreational purposes in lakes, ponds, and streams without competition, predation, or other interference by the undesirable fish (Lennon et al. 1970; Finlayson et al. 2000; McClay 2000). Now the most frequently reported uses (in order of the amount of active ingredient used) are (1) control of undesirable fish to support recreational fisheries, (2) eradication of exotic fish, (3) eradication of competing fish species in rearing facilities or ponds, (4) quantification of populations of aquatic organisms, (5) treatment of drainages before initial reservoir impoundment, (6) eradication of fish to control disease, and (7) restoration of native, threatened and endangered species (McClay 2000; 2005).

Rotenone has several advantages for obtaining control of certain fish populations over other control techniques (Table 1.1). Eradication using piscicides was more successful than control efforts for improving desirable aspects of a fishery (Meronek et al. 1996). Adjustments in rotenone applications can result in spatially selective eradications. Rotenone can be used in large river systems to control all post-embryonic life stages, and the results are nearly immediate. There are several disadvantages to rotenone use: (1) temporary loss of potable water supplies and recreational opportunities, (2) temporary effects on aquatic habitat and nontarget species, and (3) rotenone does not kill fish eggs.

Methods other than piscicides for reducing or controlling fish communities include (1) modification of angling regulations (modifications to promote or favor overharvest), (2) physical removal techniques (nets, traps, or electrofishing), (3) biological control techniques (predators, intraspecific manipulation, pathological reactions), (4) dewatering or water fluctuation techniques, (5) streamflow augmentation techniques (create water temperatures or current conditions that negatively impact the undesired species or that favor the desired species), (6) fish barriers (protect against entry by undesirable fish), and (7) explosives for flowing waters and impoundments. Major advantages of nonchemical methods are low cost and public acceptance but their collective limitation is that these techniques are usually limited to partial control, not eradication (Table 1.1).

## 1.3 Formulations and Environmental Fate

Rotenone ($C_{23}H_{22}O_6$), or (6R, 6aS, 12aS)-1,2,6,6a,12,12a-hexahydro-2-isopropenyl-8,9-dimethoxychromenyl[3,4-bfuro[2,3-h]chromen-6-one, is a botanical pesticide registered for piscicidal (fish kill) uses (Figure 1.1). Rotenone (CAS 83-79-4) is a rotenoid found in roots, seeds, and leaves of various plants that are members of the bean family *Leguminosae* from Australia, Oceania, southern Asia, and South America. Other plant rotenoids that are similarly structured to rotenone are also contained in the plants from which rotenone is extracted. Formulated end-use products of rotenone may have varying amounts of "cube root extractables" containing rotenoids. Cube resin extracted from plants contains rotenone, deguelin, rotenolone, and tephrosin (Fang and Casida 1999), and 25 other main rotenoids, but the toxicity is almost entirely due to rotenone content (Fang et al. 1997). Analytical standards are available from several research chemical suppliers (e.g., Aldrich and Sigma) and the registrants. Rotenone products are classified as Restricted Use Pesticides (RUP) due to acute inhalation, acute oral, and aquatic toxicity. Rotenone is formulated as a powder (ground-up plant root material) or extracted from plants with other rotenoids (cube resin) as liquid (with emulsifiers and solvents) for use as a piscicide (Ball 1948; McClay 2005). Rotenone powder is typically packaged in 50- to 250-pound cardboard containers with plastic liners, and rotenone liquids are typically packaged in 1-, 5-, 30- and 50-gallon plastic and metal jugs and drums. Applications are generally made with helicopters and boats in lakes, reservoirs and ponds, with direct metering into rivers and streams and with hand-held equipment such as backpack sprayers in difficult to reach areas. Rotenone may be applied at any time of year, but most applications typically occur during warm months because the compound is more effective and degrades more rapidly in warm water than cold water.

Rotenone has low to moderate mobility in soil and sediment ($K_d$ = 4–426), has a relatively low potential for bioconcentrating in aquatic organisms (BCF< 30), is not persistent in the environment (due to hydrolysis and photolysis) with half-lives measured in days and hours, respectively, and its low vapor pressure (6 x $10^{-6}$ Pa) and Henry's Law constant (estimated 1.1 x $10^{-13}$ atm-$m^3$/mol) limit volatility (Table 1.2). Confirming the low volatility of rotenone, an air monitoring study in California failed to detect rotenone in air surrounding applicators during several spray applications of liquid rotenone formulation CFT Legumine™ (Westervelt 2007). Rotenone degrades quickly through abiotic (hydrolytic and photolytic) and biological mechanisms with residues generally persisting from a day to several weeks (Finlayson et al. 2001; McMillin and Finlayson 2008).

Table 1.1. Advantages and limitations of fish removal methods (modified from Finlayson et al. 2000).

| Technique | Advantage | Limitation |
|---|---|---|
| Piscicides | • Except for dewatering, only method capable of eradication<br>• Can be applied to achieve spatially selective eradication<br>• Can be used in large river systems<br>• Rapid results<br>• Controls all post-embryonic stages | • Temporary loss of potable water supplies and recreational opportunities<br>• Temporary effects on aquatic habitat and nontarget species<br>• Does not kill fish eggs until shell ruptures in hatching |
| Physical | • Publicly acceptable | • Nearly impossible to achieve eradication<br>• Need high exploitation rates<br>• Juveniles and other game fish fill void<br>• Expensive and labor intensive<br>• Potential escapement<br>• Benefits are short term |
| Biological | • May be low cost<br>• May provide sport fishing | • Limited success in maintaining predator populations<br>• Difficulties with techniques<br>• Unpredictable results but will not result in eradication because pest species must always be present to maintain controlling organism<br>• Inability to control introduced pathogens<br>• Legal concerns<br>• Creates a clientele (fishery) for undesired or predator species |
| Dewatering | • Except for piscicides, only method capable of eradication<br>• May be low cost | • Water remains in some pools and stream sections<br>• Can be detrimental to game fish<br>• Environmentally disruptive |
| Fish Barriers | • Upstream barriers remain in place to have long-term advantages | • Not effective against downstream migration of all fish<br>• Possibly not effective under flood conditions<br>• High cost<br>• Can isolate (lack of connectivity) adjoining populations<br>• Unpopular with public |
| Explosives | • Low cost<br>• Effective in small areas | • Generally cannot eliminate entire populations<br>• Could impact dam integrity<br>• Hazardous to humans and nontarget organisms<br>• Some species are resistant |

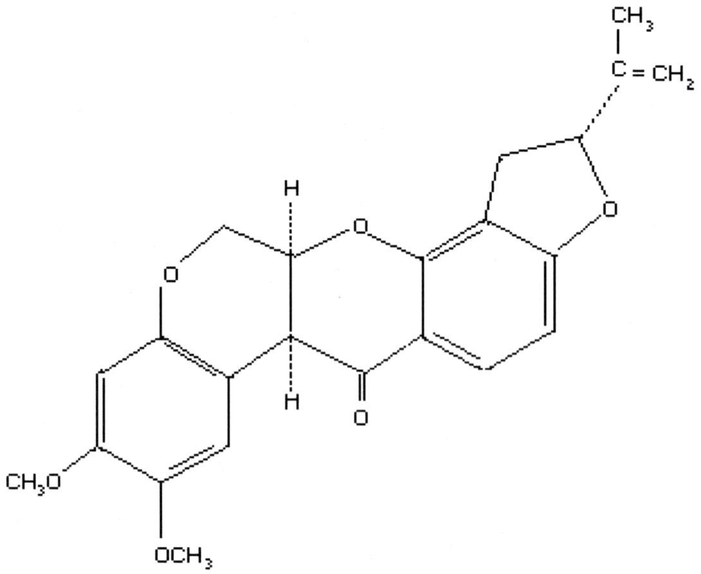

FIGURE 1.1. Structure of rotenone.

At the rotenone concentrations typically used, the target species is killed within hours of application provided that treatment concentrations are maintained at relatively constant levels. Treatment concentrations are designed to exceed the median lethal concentrations by several fold to assure complete kills of the target fish. Rotenone degrades at least moderately rapidly in aquatic environments, thus it is unlikely that residues will accumulate in water or sediment. Rotenolone is the major breakdown product of rotenone (Thomas 1983). The water treated with rotenone will generally detoxify through natural processes within one week to one month, depending upon environmental conditions. In California, Finlayson et al. (2001) and McMillin and Finlayson (2008) found half-life values for rotenone in impoundments to vary from 0.65 to 7.7 days, inversely related to temperature.

Since rotenone is applied directly to water, EPA considers the risk of terrestrial animal acute mortality to be low since there are not likely to be rotenone residues on terrestrial animal forage items. There are insufficient quantities of rotenone to represent a risk of acute effects in terrestrial animals that have consumed fish killed by rotenone or rotenone treated water (EPA 2007). The toxicity of rotenone to most terrestrial organisms is in the ppm range while that to most aquatic organisms is in the ppb range (Table 1.3). The toxicity of technical (pure) rotenone, required in standard toxicity tests for determining impacts to public health, is at least twice as toxic to animals as rotenone in commercial piscicide formulations (Table 1.3; EPA 2007). Hence, test results using technical rotenone are very conservative (effect is more severe) in predicting impacts to public health. This difference in toxicity coupled with its lack of environmental persistence, make rotenone an ideal piscicide. Since fish quickly sink to the bottom of treated water and rapidly decompose, the likelihood of chronic exposure through the diet of terrestrial animals is also considered to be low (EPA 2007).

For its use as a piscicide, rotenone is formulated as a liquid or a powder. Treatment concentrations of 25 to 200 μg/L (ppb) a.i. are recommended depending on properties of the aquatic environment and management objective. In moving water, concentrations of rotenone dissipate and decline by dispersion, dilution, hydrolysis, and photolysis and possibly induced chemical deactivation. Rotenone concentrations in standing water may persist for days to weeks and degrade by hydrolysis, photolysis, biodegradation, and partitioning to sediment (Table 1.3; Finlayson et al. 2001).

Rotenone is practically non-toxic to honeybees on an acute contact exposure basis, is slightly toxic to birds on an acute oral and subacute dietary exposure basis, and is highly to moderately toxic to rats on an acute oral exposure basis, depending on purity (Table 1.3.). Rotenone is isolated from plants and was

TABLE 1.2. Physical and chemical properties of technical rotenone.

| Property | Value | Reference |
|---|---|---|
| Molecular weight | 394.4 g/mol | Tomlin (1994) |
| Melting point | 157–175ºC | Huntingdon Life Sciences (2007) |
| log $K_{ow}$ | 4.10 | Hansch et al. (1995) |
| Water solubility (20°C) | 0.296 mg/L | Huntingdon Life Sciences (2007) |
| Vapor pressure (25°C) | $6 \times 10^{-6}$ Pa | Huntingdon Life Sciences (2007) |
| Henry's Law constant | $1.1 \times 10^{-13}$ atm–$m^3$/mol | EPIWIN (2004)—estimate |
| Hydrolysis $t_{½}$ (25°C) | 12.6 days (pH 5)<br>3.2 days (pH 7)<br>2.0 days (pH 9) | Thomas (1983) |
| Aqueous photolysis $t_{½}$ | 1.4 hours<br>8.2 hours | Spare (1982)<br>Draper (2002) |
| Soil photolysis $t_{½}$[a] | 2.9 hours | Cheng et al. (1972) |
| Sediment partition coefficient ($k_d$) | 4-426 | Dawson (1986) |
| BCF (fish) | 10.8 (viscera)<br>27.9 (head)<br>27.6 (whole carcass) | Gingerich and Rach (1985) |

[a]Soil photolysis data are not available. Estimated from rotenone applied to the surface of bean leaves.

TABLE 1.3. Acute and chronic toxicity of technical and formulated rotenone to a variety of animals.

| Species | Toxicity Value | Reference |
|---|---|---|
| Ring-necked pheasant | $LD_{50}$ = 1,680 mg/kg | Tucker (1968) |
| Rat (Female) | $LD_{50}$ = 39.5 mg/kg | Eiseman (1984) |
| | $LD_{50}$ = 320 mg/kg[a] | Lowe (2006a) |
| Honey bee | $LD_{50}$ => 60 µg/bee | Stevenson (1978) |
| Rainbow trout | $LC_{50}$ = 1.94 µg/L | Bills and Marking (1986) |
| Rainbow trout | MATC = 1.49 µg/L | Bills et al. (1986) |
| Cladoceran | $LC_{50}$ = 3.7 µg/L | Rach et al. (1988) |
| Cladoceran | MATC = 1.77 µg/L | Rach et al. (1988) |

[a]Toxicity of rotenone in CFT Legumine™ formulation.

routinely used in the past as an insecticide on plants; thus, adverse effects on plants are not suspected. Rotenone is highly toxic to fish on an acute exposure basis. Rotenone does not affect the nitrification, nitrogen fixation or the degradation of starch, cellulose or proteins (Steele 1982).

The direct application of 25 to 200 µg/L rotenone to freshwater environments as a piscicide is typically intended to kill all of the target species of fish in the target area. Rotenone is an acute poison with little potential for chronic toxicity as evidenced by the similarity of acute and chronic toxicity values for trout and cladocerans (Table 1.3.). Exposure of aquatic organisms outside the intended treatment area is limited through rigorous application of this manual's SOPs by trained fishery professionals. In flowing water environments, rotenone can be deactivated with potassium permanganate to prevent its movement out of the treatment areas. Biological and chemical monitoring of the treatment area and deactivation of rotenone where desirable will limit exposure of non-target organisms.

## 1.4 Public Health and Concerns

The acute toxicity profile required by FIFRA for pure rotenone is complete (Table 1.4). Technical rotenone is acutely toxic via the oral and inhalation routes of exposure (EPA Highly Toxic Category I), with female rats more sensitive than male rats, but formulated products containing cubé resin appear much less toxic. Rotenone is neither corrosive nor irritating to the skin or eye and is not a dermal sensitizer. The dermal toxicity is greater than 5,000 mg/kg. Rotenone does not easily penetrate human skin; only 0.37% of rotenone applied in CFT Legumine™ is absorbed through human skin (Swan 2007).

The chronic toxicity profile by FIFRA for rotenone is relatively complete (Table 1.4). Oral dosing studies for rotenone include an oral 90-day subchronic with dogs (Ellis et al. 1980), oral developmental toxicity with rats and mice (MacKenzie 1981; 1982), reproduction with rats (MacKenzie 1983), carcinogenicity with mice, rats and hamsters (Freudenthal 1981; Greenman et al. 1993), and combined chronic/cancer with rats (Tisdale 1985). Rotenone was negative in several *in vitro* mutagenicity assays (National Toxicology Program 1984). There are two chronic feeding studies for rats and mice (National Toxicology Program 1986). The primary route of excretion of rotenone in rats is in the feces with polar metabolites identified. Rotenone undergoes enterohepatic circulation with the tissue accumulation being low, typically less than 1% of administered doses. The most common toxic effect in animal studies from intermediate or long-term oral exposure was decreased body weight or body weight gain (EPA 2007). Rats are generally more sensitive than mice, and in both species, females are more sensitive than males to the effects on body weight. No evidence of carcinogenicity was seen in animal studies. No treatment-related structural external, visceral, or skeletal abnormalities were found in fetuses from treated females.

Rotenone was classified as Group E, no evidence of carcinogenicity in humans by the Cancer Assessment Review Committee of EPA on September 29, 1988. No evidence for carcinogenicity was seen in hamsters, mice or rats from available carcinogenicity studies. Administration of rotenone at doses up to 75 ppm (3.75 mg/kg/day) to rats for two years did not result in an increase in overall tumor incidence or increase in incidence of any specific type of tumor (Tisdale 1985).

A definitive target organ for rotenone toxicity has not been identified although it is known that rotenone uncouples oxidative phosphorylation by blocking electron transport at complex I within the mitochondria. Published literature within the past ten years indicate rotenone inhibits the activity of the mitochondrial electron transfer chain but also reproduces features of Parkinson's disease (Betarbet et al. 2000), including selective nigrostriatal dopaminergic degeneration and microglial activation (Sherer et al. 2003). These studies used intravenous and subcutaneous routes which are not relevant to humans; it is noteworthy that oral studies have not produced this pathology. Although rotenone is toxic to the nervous system of insects, fish, and birds, commercial products have presented little hazard to humans over many decades (Reigart and Roberts 1999). In his chapter on inhibitors of oxidative phosphorylation (including rotenone), Hollingworth (2001) does not consider rotenone a cause of Parkinson's disease. More recently, Rojo et al. (2007) found that mice and rats subjected to chronic inhalation of rotenone were asymptomatic for Parkinson's disease, and the amount of rotenone that might reach the brain through the nasal route appeared insufficient to produce a significant neuron loss. A review of the published data since the initial study by Betarbet et al. (2000) suggests that their rotenone-treated rats' model is based on atypical Parkinsonism rather than idiopathic Parkinson's disease, and that such studies are not applicable to the application of piscicidal rotenone (Höglinger et al. 2006).

Occupational and non-occupational exposure to rotenone from liquid formulations is limited to short-term and intermediate-term exposure from two pathways, oral and dermal; exposure to rotenone from powders includes the inhalation pathway. Chronic exposure is unlikely but possible from drinking water (oral). Adverse effects are unlikely given the rapid degradation of rotenone. Chronic exposure through food is not expected because of rotenone's rapid degradation and low propensity to bioaccumulate in fish.

The AFS FMCS *recommends* close interaction between the public and the natural resource agency using rotenone to ensure that public concerns are adequately addressed prior to initiating the project. Public involvement should begin at the most elemental stages of a project, typically at the development of a species-specific or water body-specific fish management plan. Rotenone is a tool used in fish management, and it is difficult to gain support for a project when there is no support for the management plan that requires rotenone treatments. Public acceptance can be gained through a variety of processes, including building consensus, accommodating concerns or by earning their consent. A successful public engagement process includes demonstrating to the public that (1) there is a problem, (2) the natural resources agency is the one to solve it, (3) the agency's approach is reasonable, (4) alternatives have been sufficiently vetted and considered, and (5) the agency is listening and responding to the public comments. The process may require facilitation.

Over the past several decades, rotenone use has been temporarily prohibited or limited in several states and provinces (California, Idaho, Montana, Michigan, Oregon and New York). These actions were initiated by different entities including the Center for Biological Diversity, Pacific Environmental Law Center, and Californians for Alternatives to Toxic Substances and the California Regional Water Quality Control Board, Lahontan Region. Cancellation of rotenone projects have placed federally-listed species at continued risk of extinction (Finlayson et al. 2005). Frequently, the lack of public acceptance for using chemicals in water and killing fish is at the root of the challenges mounted by opposition groups that become organized, secure funding, and mount legal challenges. Future uses of rotenone, even for small projects, are now threatened in several states and provinces. Often, small projects generate the greatest controversy. The controversy in many cases originates from the lack of public understanding on management decisions, purpose(s) and environmental tradeoffs associated with the project. Information on these elements will usually increase public understanding and yield a greater acceptance for the use of chemicals and the killing of fish.

In a 1998 survey of natural resources agencies in the United States and Canada, many agencies reported on issues associated with the use of rotenone (McClay 2000; Finlayson et al. 2000). Agencies overwhelmingly identified public acceptance and understanding, environmental concerns, and the "usability" of the product as the most important issues confronting them. Specifically, agencies requested information and guidance on the following broad categories (in order of frequency mentioned): (1) collection and disposal of dead fish; (2) impact of rotenone and other ingredients on public health; (3) impact of rotenone and the other ingredients on surface and groundwater quality; (4) adequate public notification and education; (5) impact of rotenone on animal welfare—fish; (6) impact of rotenone on animal welfare— wildlife; (7) impact of rotenone on invertebrates; (8) rotenone residues in fish; (9) liability and property damage; and (10) impact of rotenone and other ingredients on air quality. At least five factors may influence the public's interest in a rotenone project: (1) proximity to and size of population centers, (2) degree of public trust associated with the agency proposing the project, (3) size of treatment, (4) degree and type of other uses associated with project water, and (5) other agency management plans affected by the project.

A Public Information Program is crucial to informing the public on the benefits and impacts of rotenone use; however, dispelling fears may not always be possible. As more demands are placed on the continent's bodies of water and the public becomes more environmentally aware, there will be a need to respond with information on how rotenone is being used in a manner to minimize environmental impacts. The revised

Table 1.4. Acute, subchronic, chronic, developmental, and reproductive toxicity profile on technical (and formulated where noted) rotenone (modified from EPA 2007).

| Species | Toxicity Value | Reference |
|---|---|---|
| Acute oral (rat) | LD50 = 102 (M) 39.5 (F) mg/kg<br>LD50 = 320 mg/kg (F)[a] | Eiseman 1984<br>Lowe 2006a |
| Acute dermal (rabbit) | LD50 = > 5,000 mg/kg | Gabriel 1996 |
| Acute inhalation (rat) | LD50 = 0.024 (M) 0.019 (F) mg/L<br>LD50 > 0.062 (M&F)[a] | Hobert 1995<br>Lowe 2006b |
| Acute dermal irritation (rabbit) | PIS 0.08 at 1 hr | Moore 1995a |
| Acute eye irritation (rabbit) | PIS 3.3 at 1 hr | Moore 1995b |
| Skin sensitization | Not a dermal sensitizer | Kuhn 1995 |
| 90-day oral (dog) | NOAEL = 0.4 mg/kg/day<br>LOAEL = 2 mg/kg/day | Ellis et al. 1980 |
| Developmental (rat) | Mat NOAEL = not determined<br>Mat LOAEL = 0.75 mg/kg/day<br>Develop NOAEL = 3 mg/kg/day<br>Develop LOAEL = 6 mg/kg/day | MacKenzie 1982 |
| Developmental (mouse) | Mat NOAEL = 15 mg/kg/day<br>Mat LOAEL = 24 mg/kg/day<br>Develop NOAEL = 15 mg/kg/day<br>Develop LOAEL = 24 mg/kg/day | MacKenzie 1981 |
| Reproduction (rat) | Parental NOAEL = 0.5/0.6 mg/kg/day<br>Parental LOAEL = 2.4/3.0 mg/kg/day<br>Repro NOAEL = 2.4/3.0 mg/kg/day<br>Repro LOAEL = 4.8/6.2 mg/kg/day<br>Offspring NOAEL = 0.5/0.6 mg/kg/day<br>Offspring LOAEL = 2.4/3.0 mg/kg/day | MacKenzie 1983 |
| Chronic/Oncogenicity (rat) | NOAEL = 0.375 mg/kg/day<br>LOAEL = 1.88 mg/kg/day<br>No evidence of carcinogenicity | Tisdale 1985 |
| Chronic/Carcinogenicity (mouse) | NOAEL = <111/124 mg/kg/day<br>LOAEL = 111/124 mg/kg/day<br>No evidence of carcinogenicity | National Toxicology Program 1986 |
| Carcinogenicity (hamster) | NOAEL = 42 mg/kg/day<br>LOAEL = 83 mg/kg/day | Freudenthal 1981 |

TABLE 1.4. Continued.

| Species | Toxicity Value | Reference |
|---|---|---|
| Gene mutation (*S. typhimurium*) | No evidence of induced mutant | Haworth 1978 |
| Gene mutation (mouse lymphoma) | Evidence of concentrated-related response of induced mutant colonies w/o metabolic activation | National Toxicology Program 1984 |
| Micronucleus (mouse) | Negative at doses to 80 mg/kg/day | Biotech 1981 |

[a]Rotenone from CFT Legumine™ formulation.

labels and procedures in this *Rotenone SOP Manual* reflect mitigation measures that lower risks identified in the assessment to human health, recreational, occupational and ecological risks found in the RED (EPA 2007).

The two most important concepts that natural resources agencies should incorporate into rotenone projects are (1) public acceptance can come from public understanding, and (2) public input minimizes controversy. Involving the public early in the decision making process is crucial to public understanding and acceptance. Informing the public of a project likely will require a public meeting where a brief narrative on the project is presented. The narrative should focus on alternatives to correcting the fish management problem, nontechnical information on rotenone, explanation of project and schedule, and anticipated benefits of the project. The public will likely be interested in the project if they attend the meeting and the agency should be a good listener and ask for their comments. Agencies should strive to accurately and clearly communicate project objectives and environmental trade-offs to the public. Public support for renovating a fish community may be generated when managers can demonstrate that the current community is the result of human-induced perturbations and that the preferred alternative is complete renovation. The public often does not understand that some short-term losses may be offset by long-term benefits such as native fish restoration, improved habitat or many years of improved angling opportunity.

## 1.5 Rotenone SOP Manual Structure, Updates and Terminology

This manual is divided into four chapters: Introduction, Rotenone Project Planning Procedures, Standard Operating Procedures, and References. The first two chapters give the fishery manager background information on rotenone's (a) attributes and historical use, (b) environmental fate and behavior, (c) safety to the environment, non-target organisms, and humans, (d) reregistration history, and (e) project planning procedures. The techniques for using rotenone are outlined in 16 SOPs located in Chapter 3. The information contained in each SOP in the *Rotenone SOP Manual* generally follows this order (1) applicability, (2) purpose, (3) location on label, and (4) procedure. The final chapter contains a complete set of references. SOPs may be revised and new SOPs may be added in the future. Applicators should check the AFS website frequently for changes.

Updates will be provided as needed by the members of the AFS FMCS at www.fisheries.org/units/rotenone. The reference to the revision will be as follows: SOP. #.#., version #., month/day/year. Readers can submit comments to the AFS webpage.

The words "must," "should," "may," "can," and "might" have very specific meanings in this manual:

" Must" is used to express an absolute (mandatory) requirement, that is, to state that the guidelines are designed to satisfy the specified condition. "Must" is only used in conjunction with factors that directly relate to the legality or acceptability of specific recommendations (i.e., a requirement on the label of a pesticide product).

" Should" is used to state that the specified condition is recommended (advisory) and ought to be met, if possible. Terms such as "is desirable," "is often desirable," and "might be desirable" are used in connection with less important factors.

" May" is used to mean "is (are) allowed to."

" Can" is used to mean "is (are) able to."

" Might" is used to mean "could possibly." "Might" is never used as a synonym for either "may" or "can."

# 2 Rotenone Project Planning Procedures

A rotenone project usually will have five stages: (1) preliminary planning, where the project concept and alternatives are developed, public input is invited, and acceptance is encouraged; (2) intermediate planning, incorporating an EA where the scope of the project is refined; (3) final planning and project implementation, involving development of project-specific work plans to accomplish the application; (4) performing the treatment; and (5) summation and critique of the project into a final report (Figure 2.1).

A small treatment performed on private land or a government-owned hatchery may require little planning before implementation, while a large project involving a public water supply may require two or more years of extensive planning and discussions and conflict resolution. The rotenone treatment should be consistent with and supported by the current FMP when applicable. The complexity of a rotenone project depends upon social, biological, political, and physical characteristics and will dictate the degree of planning required.

## 2.1 Preliminary Planning

Preliminary planning is critical to the success of fish reclamation and sampling projects using rotenone. The project plan should be based on facts and tactics that firmly stand throughout the whole process. Key ingredients in preliminary planning usually include: (1) public involvement; (2) FMP; (3) statement of need; (4) determination of applicable laws and regulations; and (5) internal agency review and approval. Once these elements have been completed, an outline of a preliminary treatment plan is usually beneficial to define the scope of the rotenone project for future planning.

### 2.1.1 Public involvement

The public needs to be involved in the rotenone project, beginning with the development of the FMP at the preliminary planning stage. Individual states may have administrative procedures that must be followed in regard to holding public meetings. The public should continue to be involved as the project is developed and executed. The public may have concerns with the use of chemicals and the killing of fish (see Section 1.4). These concerns can result from the public's lack of understanding or disagreement with the project's purpose(s), environmental tradeoffs created by the project, and the management decisions leading to the project. At the completion of preliminary planning, a Public Involvement Plan should be in place to insure a formal process for public input.

A Public Involvement Plan should (1) identify each milestone for public involvement including dates for initial public notice, public meetings, written comments, final decisions, and notifications (in short a road map for the project), (2) identify key interest individuals, groups and agencies, (3) enlist key support groups and allies and assign contact persons, (4) notify news media and assign contact persons, (5) anticipate responses from individuals, groups and agencies and concentrate efforts on those likely to oppose, and (6) assess methods to inform and obtain public comment to assure the public is informed and their input is received and assessed. Most of the public should be convinced that the project is a reasonable means of correcting the conflict with the goals of the FMP.

Informing the public of the project usually involves a public meeting where the project is presented with an emphasis on the problem, and the agency asks for comments. The public is interested and involved in the project as indicated by their attendance. A brief narrative is usually prepared for the public that lists alternatives to correcting the problem, nontechnical information on rotenone, an explanation and schedule for the proposed project, and anticipated benefits to the resource and the public.

### 2.1.2 Fisheries management plan

Fisheries resources can be managed for a particular species or water body, or both. A FMP assesses the status of a specific water body or populations of specific fish species and determines the appropriate management actions necessary to maintain the desired fishery. A current FMP ensures that written goals and objectives for a specified time period are clearly defined and implemented. The type of management desired will determine which type of FMP is used. Public input during the development of the FMP is beneficial. A FMP may be inadequate for dealing with an emergency situation such as the discovery of an invasive species that must be eradicated quickly.

A species-specific FMP minimally contains the following items: (1) goals and objectives of the FMP; (2) historic habitat range of managed species; (3) description of environmental and human problems; (4) identification and prioritization of suitable habitat locations, including threats to each habitat type; (5) plan description; and (6) time line of management measures.

A FMP for a specific water body is similar to the species-specific plan except that this plan includes a description of the biotic diversity of the water body instead of a habitat range for the species. The water body specific FMP should contain the following items: (1) general geographical setting of the area; (2) description of existing land management surrounding the water body; (3) water quality and development surrounding the water body (e.g., forest, residential, industrial); (4) recreational facilities and activities; (5) ownership; (6) hydrology; (7) aquatic animal assemblages and habitat types; (8) threatened and endangered species; (9) fishery description; (10) current fish management; and (11) proposed management program (objectives, direction, recommendations).

### 2.1.3 Statement of need

A statement of need for a rotenone project should be supported by a logical progression from a FMP or other ecosystem plan. The need for a project should be clearly supported by factual evidence. For example, the presence of undesirable fish as defined by the fisheries manager may suggest rotenone treatment. The objective of the project is to correct existing fishery conditions that conflict with the goals of the FMP. Project objectives might include (1) reversing unacceptable declines in population size or growth rate of a desirable fish species, (2) eliminating undesirable species, (3) minimizing outbreaks of contagious disease, (4) reintroducing a native species into a historical range, or (5) changing the desired species composition in response to public demand. The justification should also include consideration of prevailing on-site regulations and management plans, the current and potential demand for fishing within the water body, the need to protect nearby waters from undesirable species, or the uniqueness of a remote, native fishery. Projects that emphasize single species management should address issues of ecosystem diversity. Knowledge of the presence of parks or other public facilities (or possible future developments), and the proximity of population centers is useful for this process. The justification should also explain why other options would not accomplish the desired results. If an environmental analysis is planned (see Section 2.2.2), the draft document should evaluate all realistic options or alternatives. Small projects that do not require an EA should include a clear justification for the decision to use rotenone.

Before proposing a project, a biological survey of fish community composition is necessary. At a minimum, sample a variety of species and age-groups to determine the presence of the most rotenone-resistant species subject to removal. The justification may contain measures of angler success and use, such as creel survey information and fish stocking information. Written or oral comments solicited from the angling public can provide information about general satisfaction with a fishery. The justification should include a description of the fish community, desired fish management objectives, life history of the target fish species, and a comparison of the available alternative control measures. The project might also have population estimation and enumeration as an objective. The appropriate uses of rotenone and alternative control measures are discussed in Section 1.2.

### 2.1.4 Determination of applicable laws and regulations

State the legal authorization (federal, state, or provincial) for fish and wildlife agency management of aquatic resources. Documentation of this authority may prove instrumental in countering legal challenges to the project and in negotiations with other parties. These mandates usually address the conservation, maintenance, and utilization of natural resources to ensure the continued existence of all species and the maintenance of a sufficient resource to support reasonable recreational fisheries. The natural resources agency may have specific powers to take any species which is (1) unduly preying upon a desirable species of bird, mammal, or fish, (2) an introduced species, or (3) harboring a highly contagious disease.

Determine those regulatory agencies that have overlapping jurisdictions for regulating a treatment. Agencies that regulate the following areas may require notifications, applications, approvals, and permits: (1) agriculture; (2) water use; (3) environmental protection; (4) water quality; (5) public health; (6) land use; and 7) threatened or endangered species. Determine the applicable regulations and restrictions, and obtain clearances in sufficient time before the treatment. Outside agencies may need monitoring plans and other requirements before treatment, so allow sufficient time for compliance. Resolve conflicts over regulatory and jurisdictional issues before treatment. Agreements that delineate interagency authorities, responsibilities, procedures, and time lines have been instrumental in resolving conflicts among agencies with overlapping responsibilities.

### 2.1.5 Internal review and approval

The project should receive internal review and approval using the agency's chain of command structure prior to receiving public input. It may be beneficial to use a predetermined format for assessing proposals that provide for written information on (1) ownership and use of the water body, (2) interested parties likely affected by the action, (3) environmental compliance issues, (4) water uses affected by project, (5) target species and conflicts with FMP, (6) supporting and regulatory agencies affected, (7) likely supporting and opposing groups and individuals, (8) physical and chemical characteristics of the water body, (9) chance of success and number of treatments/years required for success, (10) threat of target species to surrounding waters, (11) chance for eradication, (12) downstream areas affected by project, and (13) sensitive, endemic and listed species impacted by project.

### 2.1.6 Preliminary treatment plan

The preliminary treatment plan is the first snapshot of the proposed application from beginning to end. It needs to be completed, reviewed, and tentatively approved internally before project implementation. The preliminary plan should contain the following elements to gain an accurate assessment of necessary resources.

#### *2.1.6.1 Physical and chemical characteristics of the water body*

Prepare a general location and morphological map of the system to be treated and describe the important environmental attributes that need consideration. These may include volume or flow of water, type and density of aquatic vegetation, depth of lake, shoreline configuration and substrate, inlet and outlet flows, flushing rate, temperature, pH, dissolved oxygen, turbidity, and conductivity of water at anticipated time of treatment.

#### *2.1.6.2 Barriers, ownership, and obstructions*

Indicate and describe the barriers and obstructions to fish (hydrological) and human (topographical and legal) movement on maps. Include ownership of surrounding land and the extent and location of swampy areas and other areas that may require special treatment.

#### 2.1.6.3 *Rotenone and deactivating chemical*

Describe the type of formulation, concentrations, and amounts of rotenone likely to be used. The rotenone concentration and formulation depend on a bioassay with the target species in site water, depth, volume, water clarity, flushing rate, pH, and water temperature at the time of the proposed treatment. Typically, a lake or stream is divided into treatment zones, each with specific requirements. Specify the application rate and the amount of rotenone needed in each treatment zone. Assess the environmental advantages and disadvantages of natural degradation and the use of deactivating agent potassium permanganate. If deactivation is required, describe the type, concentration, and location of the deactivation zone. For stream treatments, distinguish between the treatment area and project area with the latter including the deactivation zone.

#### 2.1.6.4 *Public and commercial interests*

Identify and contact public and commercial groups that use the water body, especially if it is a public or industrial water supply. It might be advantageous to involve representatives from these groups. Document ownership of the land surrounding the water body to be treated and water licenses held, particularly for inlet and outlet streams.

#### 2.1.6.5 *Interagency responsibilities*

Contact all government agencies at the local, state, provincial, or federal level that might have plans, permits, authorities, or responsibilities affected by the treatment. These include health, agriculture, parks, environment, water supply and quality, air resources or quality, and land use agencies. Include local governments (counties, cities, water reclamation districts, and conservation districts). It may be desirable to assign an agency contact person for each of these outside agencies.

#### 2.1.6.6 *Logistics and preliminary schedule*

Summarize the methods of operation, number of staff needed, timing, equipment needs (purchases and rentals), required permits and approvals, and biological and chemical monitoring required, and schedule each major milestone. Develop an outline and schedule for all actions required in the Intermediate Planning (see Section 2.2) and Project Implementation and Management (see Section 2.3) sections by selecting a proposed treatment date and working backward with reasonable completion dates for the milestones (see Figure 2.1). Allow for periodic assessments to amend the schedule.

#### 2.1.6.7 *Fish rescue and removal of fishing limits*

The prospect of wasting fish in a treatment may prompt public concern. Consider the viability of a pretreatment salvage operation to allow the public to remove fish for their own use. Liberalization of fishing regulations can effectively address these concerns and improve public support. Ensure enough lead-time to implement regulatory changes. An alternative may be to rescue the desirable fish from the proposed treatment area for transplanting into another water body or for holding at a facility to restock once the treated water body can again support fish. These rescue alternatives are usually expensive but can be good public relations tools, especially if the public gets involved.

#### 2.1.6.8 *Restocking*

Most, but not all rotenone projects will be followed with an effort to restock the treated water body. Develop a restocking plan based on the proposed treatment date, management objectives, and expected rotenone degradation or deactivation time. The restocking effort should be consistent with the current FMP. Public resistance to a treatment may be due to an unwillingness to accept lost fishing opportunities or the loss of quality-sized fish. In such cases, there may be a demand for immediate stocking of large catchable fish. If such a demand can be anticipated in advance, it gives managers time to make arrangements with hatcheries to meet these needs.

##### *2.1.6.9 Personnel and equipment needs*

Determine personnel and equipment needs for pretreatment, treatment, and post treatment activities based on the type and concentration of rotenone formulation and potassium permanganate prescribed and the project logistics. This estimate should include working with the public, EA, fish rescue, monitoring, dead fish removal and disposal if needed, and fish restocking.

##### *2.1.6.10 Budget*

Determine all personnel, rotenone, potassium permanganate, legal, material, and equipment (including special items) costs for the pretreatment, treatment, and post-treatment activities of the project.

## 2.2 Intermediate Planning

This planning stage refines the preliminary project plan and clears obstacles to the treatment before Project Implementation and Management (see Section 2.3).

### 2.2.1 Environmental laws

Several federal laws that will likely affect the project include FIFRA (see SOP 2), NEPA (environmental analysis; see Section 2.2.2), CWA (waste discharge requirements; see Section 2.2.3), ESA (impacts to listed species; see Section 2.2.4), and OSHA (see SOP 2). Many states and provinces have statutes with similar restrictions that may affect the project.

### 2.2.2 Environmental analysis

An EA typically focuses on environmental impacts of the project, methods of reducing environmental damage through alternatives or mitigation, and disclosure of rationale for the project. Scheduling an EA depends on the complexity of the project and the issues involved, but the document should be completed and approved before Project Implementation and Management (see Section 2.3). Ideally, the collection of information for the EA should begin sometime during the latter stages of Preliminary Planning (see Section 2.1) to assist in refining the project scope. An EA may require one or two year's lead time, depending on the requirements of responsible agency and project issues. For projects proposed on federal lands, ensure that advance coordination with federal management agencies, and necessary approvals and documentation are completed prior to commitment of resources.

Environmental quality laws codify specific policies of federal, state, and provincial governments. These policies typically provide for (1) maintaining a quality environment, (2) identifying critical thresholds for personal health and safety, (3) encouraging systematic and concerted efforts for management of natural resources and waste disposal, (4) encouraging the enjoyment of esthetic values of natural resources, (5) preventing the elimination of fish and wildlife species, and (6) requiring government agencies to consider alternatives with lower environmental impacts.

In the United States, the NEPA (1970) sets forth a systematic approach for evaluation of the environmental impacts of federal actions, those permitted by a federal agency or those using federal funds. Many states and provinces have similar review processes, some of which tend to place a higher value on environmental protection than on economic growth or other social considerations. For proposals subject to NEPA, an agency must evaluate and consider all reasonable alternatives and must suggest appropriate mitigation measures, but is not bound to them.

Whether subject to NEPA or state or provincial environmental quality review, the normal procedure is to conduct a preliminary analysis to determine whether the proposed treatment is categorically excluded from the need for further consideration, whether to prepare an environmental assessment (under NEPA) to ascertain if the project will have no significant adverse impact on the environment or if the impact can be mitigated. Alternatively, an environmental impact statement (under NEPA) is prepared if a significant unmitigated adverse impact on the environment or an environmental change is expected.

Several states including California, Washington, and Michigan have prepared programmatic environmental studies of rotenone use in fisheries management (CDFG 1994; WDFW 1992; MDNR 1990). In California, specific rotenone projects are supported by the programmatic document. The programmatic document serves to (1) minimize redundancy on issues from project to project; (2) act as a reference for the hazards of rotenone use, treatment methods, and safety procedures; and (3) generally depict the expected impacts of rotenone use.

The draft EA should be released for public review if required before agency approval is finalized. The agency should notify the public of the draft document by placing a notice in a local newspaper of general circulation in the project locale and by sending copies to interested groups. Alternatively, a notice of the draft document can be published in the state or provincial environmental bulletin, if one is regularly published. The notice should indicate the time period for public comment, a brief description and location of the project, and how to obtain a copy of the draft document. Consider all written comments before final approval of the EA.

### 2.2.3 Waste discharge requirements

The objective of the Federal Water Pollution Control Act (1972), commonly referred to as the Clean Water Act (CWA), is to restore and maintain the chemical, physical, and biological integrity of the nation's waters by preventing point and nonpoint pollution sources, providing assistance to publicly owned treatment works for the improvement of wastewater treatment, and maintaining the integrity of wetlands. The applicator should check with the local state water quality regulatory agency to determine if NPDES permits are required or if additional restrictions are in place for the intended treatment area.

### 2.2.4 Endangered species

Depending on the location, the use of rotenone products may implicate ESA obligations for federal agencies and other persons using these products. Agencies and others should contact their local FWS Ecological Services at www.fws.gov/offices for assistance. Contact should be made approximately 6 months in advance of the planned rotenone application date since significant review may be required.

The ESA (1973) provides a means whereby the ecosystems upon which endangered species and threatened species depend may be conserved. Among other protections, ESA accomplishes this purpose through two mandates: section 7 consultation and section 9 take prohibitions. Section 7 is germane to federal agencies; in this case it applies to any federal agency applying rotenone and any federal agency providing funding to a non-federal agency for the application of rotenone. Section 9 is applicable to all persons.

Section 7(a)(2) of the ESA requires all federal agencies to insure their actions do not jeopardize existence of listed species or adversely modify or destroy their critical habitat. Actions include all activities and programs of any kind authorized, funded, or carried out, in whole or in part, by a federal agency. To ensure this section 7 mandate is fulfilled, federal agencies must follow procedures prescribed in regulation. In brief, if no listed species are present or will not be affected in any manner, no further consultation is needed. If listed species are present and "may be affected," the action agency must assess the impacts upon such species. If their biological assessment indicates listed species "may be affected but are not likely to be adversely affected" consultation may be concluded informally with written concurrence from the FWS or NOAA, the administrators of the ESA. If the action is "likely to adversely affect" listed species, formal consultation is required. The culmination of formal consultation is a written biological opinion that puts forth a "jeopardy" or "no jeopardy" conclusion. In the former, the biological opinion identifies reasonable and prudent alternatives, which must be taken to avoid jeopardy. In all cases where incidental take is likely to occur, the biological opinion includes an "Incidental Take Statement," which provides exemption for the incidental take of listed species.

Section 9 of the ESA prohibits take of listed species. Take is defined as to harass, harm, pursue, hunt, shoot, wound, kill, trap, capture, or collect, or to attempt to engage in any such conduct without a permit from the FWS or NOAA, as appropriate.

Federal agencies conducting rotenone projects in waters with threatened and endangered species need to determine whether such projects will affect listed species. The agencies may contact their local FWS Ecological Services office at www.fws.gov/offices/ for assistance in fulfilling their section 7 requirements. Non-federal entities conducting rotenone projects in waters with federally listed species, should contact their local Ecological Services office to determine whether a take permit is required prior to commencement of their work (refer to the above website for links to local offices). Contact should be made approximately 6 months in advance of the planned rotenone application dates since a significant review is required before a permit can be issued.

Federal and state agencies conducting rotenone projects in waters with threatened and endangered species will need to contact their local FWS office for direction on how they wish to permit these projects. Depending on circumstances—including existence of a federal nexus and likelihood of "affect", permission may be obtained under authority of different portions of the Act—including Sections 4 (d), 6, 7 or 10.

### 2.2.5 Public and agency issue identification and notification

Concurrent with intermediate planning is the identification of issues from the public and affected agencies that typically occur during the EA process (see Section 2.2.2). Description of these issues and resolution through alternatives and mitigation will further refine/redefine the project goals. Often, monitoring to insure that mitigation has reduced impacts and that regulatory requirements have been met result from issue identification and notification.

Federal, state, provincial, and local agencies with jurisdictions and objectives may affect the use of rotenone. Agencies with regulatory authority (discretionary approval) over rotenone use are considered responsible agencies. In addition to the fish and wildlife agency proposing the use of rotenone, these agencies may include EPA and state or provincial departments of food, agriculture or pesticide regulation, public health, natural resources (i.e., forestry, lands, and parks), water quality, and environment. The use, registration, and control of pesticides in the United States ultimately rest with EPA. State or provincial departments that regulate pesticides, food, or agriculture enforce pesticide laws and issue licenses and certificates for pest control operations. Many states require that a licensed Agricultural Pest Control Advisor make a recommendation to use rotenone. Only a licensed, qualified applicator (i.e., Certified Applicator) can supervise the application of rotenone. Agencies that regulate pesticides generally have concerns with (1) safety gear, (2) safety procedures, (3) disposal of used pesticide containers and dead fish and (4) the potential effect on nontarget species including humans.

State or provincial departments of health services often cooperate with the pesticide regulatory agencies in investigations of pesticide-related illnesses and develop employee safety standards for handling pesticides. Health agencies have also been delegated by the EPA to enforce the Federal Safe Drinking Water Act (1966 amendments) through such measures as adoption of drinking water standards and monitoring regulations. Public health concerns expressed by agencies over the use of rotenone have included (1) nuisance of flies and odors created by decaying fish, (2) consumption of dead fish containing bacteria and residues of rotenone and other compounds, (3) consumption of drinking water containing residues of rotenone and other compounds, (4) inhalation of pesticides, and (5) pesticide odors.

State or provincial departments of water quality or the environment typically regulate storage and transport of hazardous wastes, disposal sites for pesticide containers, and water quality standards. Water quality and environmental agencies may establish water quality control plans that reflect water quality objectives for specific hydrologic basins. Concerns with rotenone use from environmental agencies have included (1) impacts on beneficial uses of water, (2) maintenance of water quality standards, and (3) impacts on aquatic life other than fish. Rotenone treatments may also affect the activities and interests of other agencies such as counties, cities, water reclamation districts, irrigation districts, and other resource agencies.

Consult the management plans of water quality and environmental agencies, the FWS, FS, and other affected agencies to ensure their existing management plans are considered.

### 2.2.6 Monitoring program

It is desirable to monitor the application of rotenone to ensure that an effective treatment is achieved. Monitoring rotenone concentrations, efficacy and impacts to aquatic resources will likely limit potential litigation and aid in the assessment of aquatic resource recovery. Monitoring studies can help address public fears about the treatment. Develop monitoring programs during intermediate planning after important issues, special interest areas, and the scope of the project have been clearly identified. Monitoring is done to demonstrate compliance with environmental regulations (e.g., waste discharge requirements from a NPDES permit) and to insure that proposed mitigation measures were implemented and are effective.

Rotenone formulations contain a variety of compounds including rotenone, dispersants, and emulsifiers, and their dissipation in the environment over time may be of interest. The number and location of sample sites and sampling frequencies will vary with each treatment. Several monitoring methods have been used for rotenone treatments, including analyzing water, sediment, and air samples for residues of rotenone and other compounds and assessing the impact of the treatment on biological (e.g., fish, amphibian, and invertebrate) resources. Monitoring programs typically originate during the intermediate planning stage in response to the EA.

## 2.3 Project Implementation and Management

The last stage before treatment is to finalize plans for all operations associated with the project. A project schedule and structure are needed to organize large projects. For large, complex, or controversial projects, you may wish to employ a version of an ICS to organize the various functions. For some treatments no ICS may be needed, while others may require many of the ICS functional elements. Regardless of the structure chosen, qualified personnel (i.e., Certified Applicator) knowledgeable of rotenone and trained in planning and executing successful projects must supervise the treatment. Administrative approval for the treatment is obtained from the highest possible level in the natural resources agency commensurate with the scope of the project, ensuring adequate agency review. Assignments should be made for completing the project-specific work plans by specific dates. Each plan should contain sufficient information and detail so others can use the plan and complete the needed activity without assistance. The agency should complete all plans at least one month before treatment to allow for sufficient review and approval time. Also, the agency should ensure that all approvals from other agencies have been obtained and documented (see Section 2.1.4).

Depending on the size, complexity, and location of the treatment, plans may be needed for some or all of the following operations listed below. All of the SOPs in this manual provide information critical for successful planning, and these are referenced below with the relevant plan.

- <u>Fish rescue and removal of fishing limits plan.</u>

- <u>Rotenone application plan.</u> Two SOPs describe the use of bioassays to either help properly select treatment levels (SOP 5) or to monitor the efficacy of treatments (SOP 14). Five SOPs address specific delivery techniques and methods: the design and operation of semi-closed probe systems for application of liquid concentrate (SOP 8), operating protocols for the application of powdered rotenone (SOP 9), techniques for the application of liquid rotenone to streams, rivers, lakes and ponds (SOP 11), an operating protocol for spray application of pre-diluted liquid rotenone (SOP 12) and the preparation and use of rotenone powder/gelatin/sand mixture (SOP 13).

- Monitoring plan. SOP 16 provides guidance on monitoring requirements and analysis of water samples for rotenone concentrations.
- Site safety plan. Two SOPs address training: SOP 2 provides recommended supervisory training and important qualifications, while SOP 3 provides guidance on safety training and hazard communication to those involved in rotenone application. SOP 1 provides a protocol for posting treatment area restrictions prior to, during and following the application of rotenone. SOP 4 provides protocols for safe storage and transport of rotenone. SOP 10 provides guidance on proper and safe procedures for transferring liquid rotenone concentrate from product containers to service containers and application equipment;
- Site security plan. SOP 6 provides criteria for determining the treatment and project areas to assure that public safety and legal requirements are met;
- Fish removal and disposal plan. SOP 15 provides guidance for the collection and disposal of dead fish.
- Spill contingency plan. SOP 4 provides guidance on spill prevention and containment.
- Site deactivation plan. SOP 7 provides guidance for determining the need for deactivation and methods for applying potassium permanganate, while SOP 14 describes the use of *in-situ* bioassays to monitor efficacy of deactivation efforts.
- Communication plan. SOP 1 provides an operating protocol for public notification of treatment area restrictions prior to, during and following the application of rotenone.
- Restocking plan.
- Records maintenance plan.

## 2.4 Treatment

The planning process outlined in Sections 2.1 through 2.3 should have prepared the natural resources agency technically, politically, socially, and legally for the rotenone treatment. Adhering to the rotenone label and the guidance offered in SOPs 1 through 16 in Section 3 will assure a safe and effective treatment. Understanding the techniques for rotenone use will provide a sound foundation for planning a treatment. It is imperative that these planning activities occur before treatment. The treatment occurs at the end of the planning process. Once the treatment has begun, the success of the project-specific work plans in meeting objectives should be monitored and, if necessary, amended to achieve the necessary objectives.

### 2.4.1 Crises

A carefully planned treatment will generally not result in crises which loosely defined are adverse public reactions, excessive news media interest, or serious agency concerns with an unplanned event. Events that have produced crises during rotenone treatments in the past include:

- Deactivation failure, with fish killed outside of project area
- Pesticide odors occurred and resulted in public complaints
- Reported illnesses occurred to applicators or the public
- Rotenone formulation chemicals persisted longer than expected
- Project failed to accomplish stated objectives
- Excessive public opposition

### 2.4.2 Crisis management plan

In anticipation of a treatment not going according to the plan, it is prudent to be prepared to adjust the actions accordingly. For this reason, a crisis management plan should be developed before the treatment. This action plan prepares the agency for any negative development that may jeopardize the rotenone application or its favorable outcome. Before treatment, a crisis team should be identified that will act as an early alert group to develop the situation responses to problematic activities including those listed above and then use the appropriate crisis team participants and support groups.

The crisis team—The crisis team should include (1) early alert members (i.e., persons who can handle the crisis and devote exclusive time to the crisis), (2) primary response members (i.e., technical experts in various disciplines), and (3) secondary response members (i.e., high-level persons in the treatment agency, elected officials, and law enforcement).

Situation response—The situation response comprises the following steps: (1) define the problem and scope; (2) identify targets and issues; (3) select appropriate crisis team; (4) gather facts; and (5) identify a spokesperson.

Support groups—Support groups normally consist of members of (1) research groups, (2) sports clubs, associations, and organizations, and (3) regulatory governmental agencies. Gain support before the incident becomes a major crisis.

### 2.4.3 Step-by-step management of crises

Define the real problem—Gauge public actions and opinions, perhaps by using a newspaper clipping service. Focus on long-term consequences; do not focus on the details. Delegate details to support groups.

Identify a crisis team—Choose the team carefully for the situation from your preselected lists (early alert, primary response, and secondary response members). These individuals should devote themselves entirely to the crisis. Do not delay; act immediately!

Resist combative instincts—No matter what circumstances produced the crisis, keep control, or control of the situation will be lost.

Centralize control of information—Centralize control of information that is released to the public and keep the message consistent and clear.

Communicate and negotiate at the highest level—Follow the chain of command and brief all involved. Keep administration informed.

Contain problem quickly—Contain the problem quickly and stop the erosion of public confidence.

## 2.5 Project Critique

### 2.5.1 Short-term assessment

The agency should analyze the immediate effectiveness of the treatment and any mitigation measures. Goals for the short-term assessment include (1) determination of the effectiveness of chemical application (i.e., distribution and deactivation of rotenone), (2) determination of when the public can reenter the treatment area and resume use of water for drinking and irrigation, and (3) recovery of baseline environmental conditions before stocking fish. All personnel should be debriefed as soon as the treatment phase of the project has been completed to identify problems, determine causes, and propose corrective measures for future treatments. This effort involves the assessment of chemical and biological monitoring data and review of notes and observations recorded during and immediately following the treatment.

#### *2.5.1.1 Assess the effectiveness of treatment and deactivation*

The assessment of the effectiveness of the treatment and deactivation operations will enable project leaders to adjust plans based on the actual results. The effectiveness of these operations and related mitigation efforts can be judged by (1) counts of dead fish, (2) mortality of fish in live-cages (see SOP 14), (3) bioassays off site with treated water, (4) sampling for the presence of live fish, (5) measurement of concentrations of rotenone in treatment and deactivated areas (see SOP 16) and measurement of permanganate in deactivated area (see SOP 7), and (6) visual observations. The sampling of baseline environmental conditions and estimates of dead fish from shoreline counts (static-water treatments) or collections from block nets (flowing-water treatments) are useful in evaluating the effectiveness of the treatment and deactivation. In flowing-water treatments, block nets and live-cages placed at various intervals downstream from deactivation stations are effective in determining the point at which complete deactivation occurs and therefore the extent of the actual impact zone. Real time data on fish survival can be used to adjust treatment and detoxification rates, and real time data on permanganate residuals can be used to adjust deactivation rates.

#### *2.5.1.2 Recovery of baseline environmental conditions*

If baseline levels of nontarget species and environmental conditions were evaluated before the treatment, the agency needs to evaluate these parameters after the treatment to determine if recovery objectives were met and if mitigation measures are needed. Survival and recovery of the aquatic community may be demonstrated by sampling plankton, macroinvertebrates (aquatic insects, crustacea, leeches, and mollusks), and amphibians (frogs, tadpoles, and larval and adult salamanders).

Before restocking, the agency needs to test the receiving water to determine if rotenone has been sufficiently deactivated or has dissipated adequately to assure survival of stocked species. While chemical testing is recommended, it is also recommended that live-cages containing sensitive indicator species be used. The cages need to be placed in representative locations and in areas where test fish will not be killed by stress or some event unrelated to the treatment (e.g., vandalism, predation, temperature). The agency should be wary of embayments, bayous, or backwater or deepwater zones that may still harbor pockets of toxic water. In stratified lake environments, the cages should be placed at depths where there is adequate oxygen for test fish to survive.

#### *2.5.1.3 Written critique*

A written critique may be the most significant part of the treatment, and if properly done will assist the agency in improving planning and implementation of future projects. The agency should prepare a written summation and critique of the treatment as soon as possible after the treatment has been completed, in collaboration with agency's legal advisors. The agency should solicit input from all personnel involved in the treatment. It is advisable to have a meeting of all those involved in the treatment to get consensus on what worked and what should have been done differently. Each implemented plan from Section 2.3 should be assessed for accomplishing stated objectives. The agency should determine if the plan was followed, what problems or issues were associated with the plan, and what improvements were needed. The agency should send a draft of the written summation and critique to all personnel involved for review and consensus before completion of the final critique. When appropriate, the agency should use the critique to update policies and procedures.

### 2.5.2 Long-term assessment

The agency should analyze the long-term impact of the project by evaluating the stated objectives over time. Goals for the long-term assessment include (1) determination of the treatment effectiveness and benefits, (2) determination of the success of the mitigation measures in lessening the environmental impacts from the treatment, (3) assessment of the public perception of the success of the project, and (4) an overall assessment of the project.

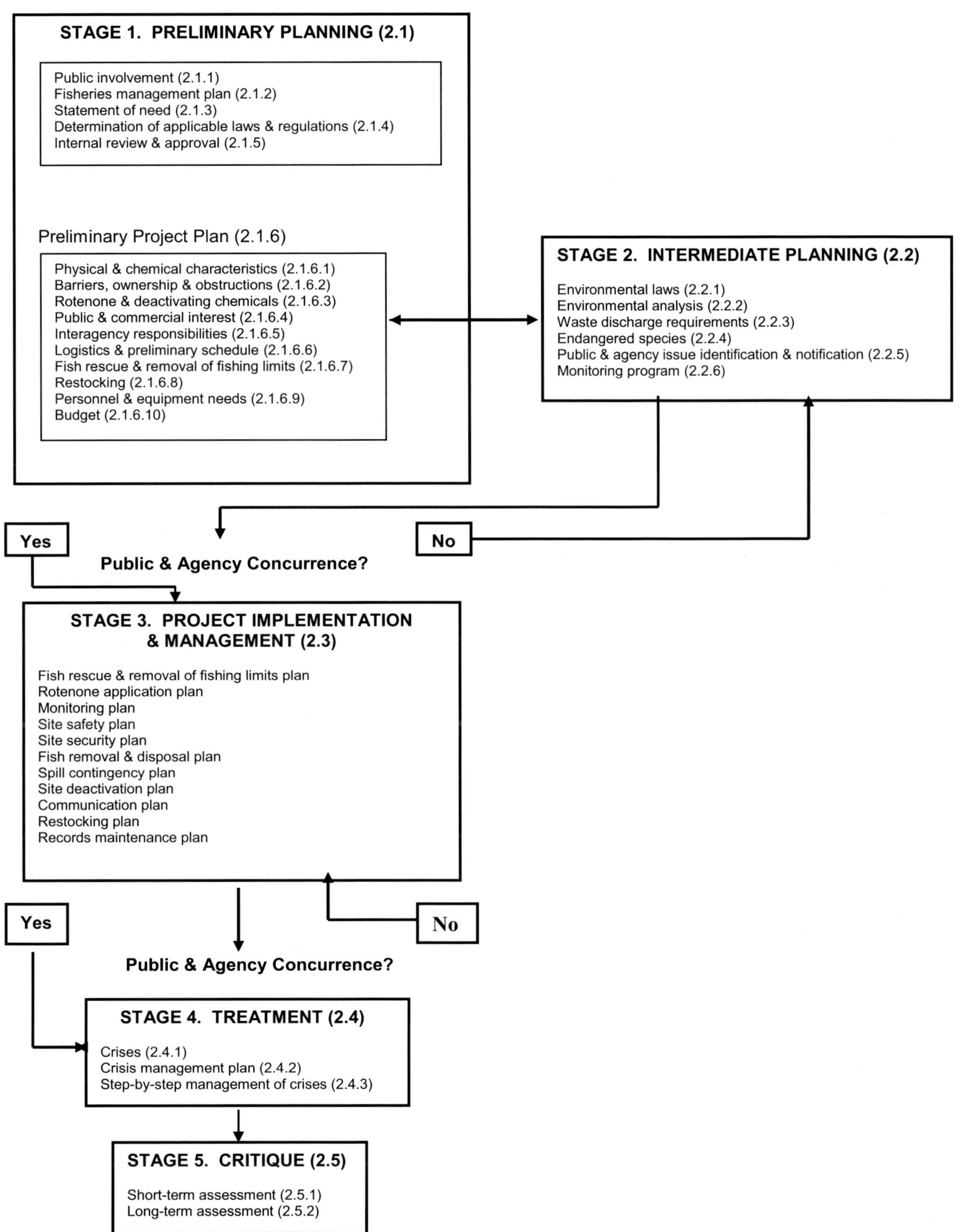


FIGURE 2.1. Five (three planning) stages of a rotenone project.

# 3 Standard Operating Procedures

As part of the reregistration process for rotenone, the EPA required a manual that contained specific procedures on how to minimize nontarget exposure and effects and to provide guidance on the new and complex label use conditions. With the approved reregistration came several significant technical changes in how rotenone will be used as a tool in fish management. These changes are incorporated into SOPs in this manual and provide guidance on how to comply with the label and use rotenone in a safe and effective manner. Many of the SOPs in the manual (SOPs 5, 6, 7, 8, 9, 10, 13 and 16) are referenced on the label and must be followed to the extent dictated by their wording. The manual contains other SOPs which, even though largely advisory, complement the safe and effective use of rotenone and should be understood and followed.

There are 16 SOPs in this section of the manual:

1. Public Notification and Treatment Area Restrictions
2. Supervisory Training and Qualifications and Regulatory Compliance
3. Safety Training and Hazard Communication
4. Rotenone Storage, Transportation, and Spill Containment
5. Determining Treatment Rates and Strategies
6. Determining Treatment Areas and Project Areas
7. Determining Need and Methods for Chemically Induced Deactivation
8. Operation of Semi-Closed Probe Systems for Application of Liquid Rotenone Concentrate
9. Operation of Semi-Closed Aspirator Systems for Application of Powdered Rotenone
10. Transferring (Mixing/Loading) Liquid Rotenone Concentrate
11. Operation of Drip Stations, Peristaltic Pumps and Propwash Venturi for Application of Liquid Rotenone
12. Operation of Sprayers for Applying Diluted Liquid Rotenone
13. Use of Rotenone Powder/Gelatin/Sand Mixture
14. Use of *In-Situ* Bioassays to Monitor Efficacy
15. Collection and Disposal of Dead Fish
16. Monitoring Requirements for Aquaculture and Drinking Water

# Public Notification and Treatment Area Restrictions
# SOP:1.0

**PROCEDURE TITLE:** Public Notification and Treatment Area Restrictions

**APPLICABILITY:** Application of rotenone in waters of the United States

**PURPOSE:**
1. Mitigate for human recreational exposure to rotenone
2. Provide an operating protocol for public notification of treatment area restrictions prior to, during, and following application of rotenone

**LOCATION ON LABEL:** "Placarding of Treatment Areas" under the heading "Directions for Use"

**PROCEDURE:**

I. Press release

A. Send general press releases to media outlets within the watershed and adjoining areas a maximum of three weeks and a minimum of one week in advance of treatment. Information includes:

1. A description of the project area
2. The reason for the treatment
3. The name and concentration of the rotenone formulation used, and deactivation agent if applicable
4. Any public or water use restrictions
5. Posting procedures
6. The anticipated length of time the area will be affected
7. The names and contact information of designated applicator and/or agency contact person(s).

B. Media outlets include newspapers and radio and television stations.

C. Press releases are provided to the media for voluntary publication or for broadcast to the public. Press releases are also valuable as an information transfer when posted at public parks and beaches, boat launch sites, and bait and sport shops around the general treatment area.

D. Provide a copy of the news release to those immediately affected by the treatment such as residents, property or business owners, and riparian users in or adjacent to the treatment area. Include those within a reasonable distance of the shoreline or stream bank affected by the rotenone treatment, including any waters treated with potassium permanganate to deactivate rotenone-treated waters. Notification by mail, email, or by handbills given directly is acceptable.

1. Property owners can be determined by accessing county records, usually by tax parcel number
2. Water rights holders can be determined by accessing records from the state or local water resources agency

II. Placarding of treatment area

A. Locations

1. The treatment area at public access points
   a. Trailheads
   b. Roads
   c. Trails

**OR**

2. Every 250 ft (76 m) along treatment area

B. Time period (access sites posted 1 day prior to treatment and application sites posted at time of application)

1. For both lotic (flowing water) and lentic (standing water) applications of ≤0.09 ppm (90 ppb) active rotenone (≤1.8 ppm 5% a.i. formulation), signs can be removed once application is complete.

2. For flowing water applications >0.09 ppm active rotenone (>1.8 ppm 5% a.i. formulation), signs can be removed following a 24-h bioassay demonstrating survival of bioassay fish, or when analytical chemistry shows ≤0.09 ppm active rotenone, or 72 hours after the application is complete, whichever is less.

3. For standing water applications >0.09 ppm active rotenone (>1.8 ppm 5% a.i. formulation), signs can be removed following a 24-h bioassay demonstrating survival of bioassay fish, or when analytical chemistry shows ≤0.09 ppm active rotenone, or 14 days after the application is complete, whichever is less.

4. See SOP 14 for bioassay techniques and SOP 16 for analytical chemistry techniques for monitoring rotenone concentrations in water.

5. Signs must remain legible during the entire posting period.

C. Required placard information (see 40 CFR for specifications)

- "DANGER/PELIGRO"
- "DO NOT ENTER WATER/NO ENTRE AGUA: Pesticide Application"
- The name of the product applied
- The purpose of the application
- The start date and time of application
- The end date and time of application
- "Recreational access (e.g., wading, swimming, boating, fishing, etc.) within the treatment area is prohibited while rotenone is being applied."
- "Do not swim or wade in treated water while placard is displayed."
- "Do not consume dead fish from treated water."
- The name, address, and telephone number of the responsible agency or entity performing the application.

D. Acceptable placard materials to withstand 14 days exposure

1. Heavy waterproof stock
2. Laminated in plastic

# Supervisory Training and Qualifications and Regulatory Compliance
# SOP:2.0

**PROCEDURE TITLE:** Supervisory Training and Qualifications and Regulatory Compliance

**APPLICABILITY:** Application of rotenone in waters of the United States

**PURPOSE:**
1. Mitigate for occupational and environmental exposures to rotenone
2. Provide recommended supervisory training and qualifications enabling the successful planning and execution of a rotenone project

**LOCATION ON LABEL:** "Hazards to Humans and Domestic Animals" under the heading "Precautionary Statements"

**PROCEDURE:**

I. Training Program for Rotenone Application

A. Certified (Qualified) Applicators supervising any aspect of the application of the project should receive training on the *Rotenone SOP Manual*. Training is available as a 4½-day course entitled *Planning & Executing Successful Rotenone and Antimycin Projects* through the American Fisheries Society at www.fisheries.org/units/rotenone/. The course is typically offered at Utah State University, Logan each May immediately prior to Memorial Day. The course has also been specially offered at other locations in cooperation with, and the request of, state fish and wildlife agencies.

B. Additionally, a project supervisor should have participated in all planning and field aspects of a minimum of two previous rotenone projects and have supervised a minimum of one aspect of a previous application prior to supervising a rotenone project.

II. Project Supervisor Qualifications and Responsibilities

A. Possess a valid Certified Applicator Certificate or license issued by respective state authority and be able to remain on site for the duration of the application.

B. Be familiar with application and deactivation equipment and techniques.

C. Ensure that the application (and employees) is in compliance with FIFRA and the product label (see FIFRA and Label Compliance below) and in compliance with OSHA (see OSHA Compliance below).

D. Provide safety training, safety training records, and personal protective equipment in a state of good repair for project personnel (see SOP 3.0).

E. Develop strategies for fish sampling/control/eradication that reflect sensitivities of target species, characteristics of rotenone, and important environmental conditions (see SOP 5.0).

F. Develop preliminary, intermediate, and implementation and management plans for public involvement, application, deactivation, monitoring, and safety.

G. Develop management and planning strategies that deal positively and effectively with unanticipated events before they occur. The resulting crises may involve the public and news media.

H. Implement application and deactivation techniques that minimize impacts.

I. Explain rotenone label and MSDS contents and requirements and how these affect use and protect human and environmental health of project personnel and the general public.

J. Characterize effects on target and nontarget organisms and environmental fate of rotenone under specific treatment conditions and communicate that to project personnel and the general public.

K. Describe key environmental laws, regulations, and processes and how these affect the specific use of rotenone to project personnel.

L. Have command of water chemistry and temperature, water travel time, discharge measurements, kinetics of deactivation, use and operation of all equipment, on-site toxicity, and safety procedures and equipment.

## III. FIFRA and Label Compliance

EPA regulates the use of pesticides under the authority of two federal statutes: the FIFRA and the FFDCA. The regulations implementing the intent of FIFRA can be found in CFR Title 40. FIFRA provides the basis for regulation, sale, distribution and use of pesticides in the United States. FIFRA authorizes EPA to review and register pesticides for specified uses. EPA also has the authority to suspend or cancel the registration of a pesticide if subsequent information shows that continued use would pose unreasonable risks. When a material is registered as a pesticide there is a presumption that there are no unreasonable risks to humans or the environment associated with the specified use. FFDCA authorizes EPA to set maximum residue levels, or tolerances, for pesticides used in or on foods or animal feed.

No one may sell, distribute, or use a pesticide unless it is registered by the EPA. The EPA must classify each pesticide as either "general use," "restricted use," or both. "General use" pesticides may be applied by anyone, but "restricted use" pesticides may only be applied by Certified Applicators or persons working under the direct supervision of a Certified Applicator. Applicators are certified by the state under a certification program approved by the EPA. All rotenone products are Restricted Use Pesticides and can only be applied by Certified Applicators trained in aquatic pest control, or persons under their direct supervision.

All pesticide label language must be approved by EPA before a pesticide can be sold or distributed in the United States. The rotenone labels contain critical information that should be discussed with those applying the material including:

- Brand name and EPA registration and establishment numbers
- Restricted use statement (rotenone is a Restricted Use Pesticide)
- Ingredient statement
- Pesticide hazard class (signal word) and first aid instructions
- Precautionary statements and PPE requirements
- Instructions for storage and disposal
- Registrant name and address
- Net weight or volume of container
- Instructions for use

The rotenone label provides instructions for its safe and effective use that should be discussed with those applying the material including:

- Treatment site
- Treatment concentration
- Treatment method
- Dilution instructions
- Treatment timing and frequency
- Re-entry interval

The overall intent of the label is to provide clear directions for effective product performance while minimizing risks to human health and the environment. It is a violation of federal law to use a pesticide in a manner inconsistent with its labeling. The courts consider a label to be a legal document. In addition, following labeling instructions carefully and precisely is necessary to ensure safe and efficacious use.

It is important that the user carefully study and fully understand the label. When in doubt, the user should contact the manufacturer or their state pesticide regulatory agency. The *Rotenone SOP Manual* or an expert associated with the AFS (www.fisheries.org) can provide guidance, but these are not legal authorities. The SOPs in this manual provide guidance on a variety of procedures for storage and spill containment of rotenone products, determining appropriate treatment rates and strategies, operation of application systems for rotenone products, and monitoring and deactivation. Individuals applying rotenone must do so in a manner not only consistent with federal laws, but also consistent with state laws and regulations, which differ from state to state. Additionally, the agency with primary responsibility for regulating pesticide use differs in each state. It is important to consult with the state pesticide regulatory agency to determine:

- If a particular rotenone product is registered for use in that state
- Rules and regulations governing pesticide use in that state
- Notifications or postings prior to pesticide applications
- How to become a certified pesticide applicator

The rotenone label must be available at each mixing and application site and it would be prudent and advisable to have a copy of the *Rotenone SOP Manual* available at each site as well.

## IV. OSHA Compliance

OSHA regulates the safe and healthful working conditions for working men and women in the United States under the authority of the Occupational Safety and Health Act (CFR Title 29). In meeting this goal, OSHA has many rules and regulations which direct rotenone handling and storage. Among OSHA directives is the HCS which is based upon the premise that employees have both a need and a right to know the hazards and identities of the chemicals to which they are exposed. They also need to know what protective measures are available to prevent adverse effects from occurring. The HCS is designed to provide employees with the information they need. The HCS requires each employer to have a CHCP and to provide their employees with communication and training relative to all hazardous materials which they may encounter in their work place. MSDS, along with training and the product label are integral parts of the CHCP.

OSHA requires all rotenone registrants to obtain or develop a MSDS for each hazardous chemical they produce or import. The data sheets include safety information generally applicable for safe handling and use which are known to the rotenone registrant and importer preparing the MSDS, including appropriate hygienic practices, protective measures during repair and maintenance of contaminated equipment, and procedures for clean-up of spills and leaks. They also include appropriate engineering controls, work practices, and PPE. Employers are required to maintain in the workplace copies of the required MSDS for each hazardous chemical and ensure that they are readily accessible to employees.

MSDSs contain vital information that is important not only to rotenone handlers but also first responders in case of a spill or fire. It is imperative that all are familiar with the MSDS information and have it available at all times. The information contained in an MSDS includes:

- Chemical or product identification
- Manufacturer's name and contact information
- Hazardous components by chemical and common names and safe exposure limits
- Physical and chemical characteristics
- Fire and explosion hazard data
- Reactivity data (incompatibility and decomposition)
- Health hazard data and first aid for four routes of exposure
- Precautions for safe handling and use
- Control measures including PPE

As mentioned above, employers are required to have a CHCP to identify all chemical hazards to the employees. Employers are also responsible for instructing employees on how to handle and use pesticides and on how to comply with pesticide laws and regulations (see SOP 3).

As a general rule, an applicator must follow the product label when applying rotenone and should follow the product MSDS for all other storage and handling procedures. Although there are OSHA requirements for the content of MSDSs and employers are required to communicate hazards to their employees using the CHCP, the rotenone product label is approved by EPA and must also be adhered to. Rotenone product MSDSs are not approved by OSHA or EPA. Employers are required by OSHA to keep records of any training. Consult with OSHA, the state OSHA or safety officer for more information.

# Safety Training and Hazard Communication
# SOP:3.0

| | |
|---|---|
| **PROCEDURE TITLE:** | Safety Training and Hazard Communication |
| **APPLICABILITY:** | Application of rotenone in waters of the United States |
| **PURPOSE:** | 1. Mitigate for occupational exposure to rotenone<br>2. Provide safety training and hazard communication to mixers, loaders, applicators and others involved with the application of rotenone |
| **LOCATION ON LABEL:** | "Hazards to Humans and Domestic Animals" under the heading "Precautionary Statements" |

**PROCEDURE:**

I. Safety Training

Employees involved in a rotenone application must be trained annually on how to safely use rotenone as explained in this SOP. Employees are given this training before working with rotenone. Rotenone is a Restricted Use Pesticide that can only be used by or under the direct supervision of a Certified Applicator. Employees wearing respirators for powdered rotenone require extra training and medical certification that they are capable of wearing a respirator without causing physical stress. All the information in the training must be in written form and employees must sign a written record verifying the training as indicated in a Record of Pesticide Training (see Appendix A for example).

The training is based on Hazard Communication Standard (29 CFR 1910.1200) and on the premise that employees have the need and the right to know the hazards and identities of the chemicals they are exposed to while working on a rotenone treatment. Employers are responsible for employees handling and using pesticides in a manner consistent with labeling, laws, and regulations and need to keep records of certain written materials (see Table SOP 3.1).

II. Comprehensive Hazard Communication Plan (CHCP)

The Hazard Communication Standard or CHCP for rotenone normally consists of (1) verbal warning, (2) review of the rotenone product label and SOPs, (3) review of the rotenone product MSDS, and (4) instruction on application of rotenone. Specifically, training must include (1) potential health effects of rotenone, (2) what to do in an emergency and emergency care available, (3) personal protective equipment required for rotenone, (4) how to use rotenone safely, and (5) rights as an employee and where to find out more information on rotenone. An example of a Hazard Communication Standard training presentation for rotenone can be found in Appendix B. Specific information is found on the MSDS and the label of the rotenone product and general information is found in the California Department of Pesticide Regulation Pesticide Safety

Information Series (see Section IV). Follow safety and PPE recommendations on the rotenone label when applying the product and the MSDS when storing or transporting the product.

A. Verbal Warning

A verbal discussion of the information and warnings on the label and MSDS for the rotenone product will suffice as a verbal warning.

B. Review of Rotenone Product Label

When reviewing the product label and SOPs, the following information should be discussed with employees: (1) what chemicals are in the rotenone product, (2) first aid and health warnings, (3) proper use of protective equipment required, and (4) directions for applying rotenone. Following the label will result in a safe mixing and application of rotenone. The label must be at the place where the rotenone is mixed and applied (see Table SOP 3.1).

Specifically, the directions for applying rotenone should provide instruction on (1) application site, (2) dosage rate, (3) application method and equipment, (4) dilution instructions, (5) application timing and frequency, and (6) restricted entry interval.

C. Review of Rotenone Product MSDS

When reviewing the rotenone product MSDS, the following information should be discussed with employees (1) health effects, (2) what to do in an emergency, (3) personal protective equipment, (4) pesticide safety, and (5) rights of employees and where to find out more information on rotenone.

1. Health Effects—Information on how rotenone can affect health is found in the Hazards Identification Section and in Toxicological Information Section of MSDS.

2. What To do in an Emergency—Information on first aid and where to get emergency medical care is found in the First Aid Measures Section of MSDS.

3. Personal Protective Equipment—Information on the need to wear PPE, how to take care of PPE and what PPE can and cannot protect is found in Exposure Controls/Personal Protection Section of MSDS.

4. Pesticide Safety—Information on the meaning of safety statements and safety rules for handling pesticides (e.g., Pesticide Information Series listed under Additional Information).

5. Rights as an Employee and More Information—Job safety information, safety leaflets, MSDS and MSDS Pocket Dictionary informs the employee about the pesticide and its dangers. Each employee has the right to know when and where the pesticide was applied, the name of the pesticide, and the EPA registration number.

D. Instruction on Application of Rotenone

Review how all application equipment works, application timing and calibration, and the proper use of personal protective equipment on the product label.

TABLE SOP 3.1. Summary Table of Records Employer Must Keep.

| Information | Storage Location |
|---|---|
| Record of pesticide training | Employer's office site |
| Written training program | Employer's office site |
| Respirator program procedures | Employer's office site |
| Pesticide label | Work site |
| Pesticide Safety Information Series | Employer's office site |
| Material Safety Data Sheet | Employer's office site |
| Storage area posting | Storage area |
| Emergency medical care notice | Work site |
| Doctor's report for respirator use | Employer's office site |
| Pesticide use records | Employer's office site |

## III. Additional Information

California Department of Pesticide Regulation. N No. 1 (Working Safely with Pesticides in Non-Agricultural Settings. Available: www.cdpr.ca.gov/docs/whs/pdf/hs1742.pdf).

California Department of Pesticide Regulation. N No. 8 (Safety Rules for Pesticide Handlers in Non-Agricultural Settings. Available: www.cdpr.ca.gov/docs/whs/pdf/hs1749.pdf).

# Appendix A

**Rotenone Product ____________________ [1]Safety Training Record**

This is to certify that __________________________________has received _________hours of rotenone safety training.

Employee Signature_______________________________
Employee Title____________________________________

Employee Work Location___________________________Trainer___________________________
Date______________

**Training Requirements:**

**Safety Procedures:** personal protective equipment, engineering controls, and equipment. Heat-related illness
**Pesticide Labels:** signal words, precautionary statements, first aid instructions, mixing and application instructions
**Pesticide Handling Procedures:** container handlings, mixing and application equipment, triple rinse containers
**First Aid and Decontamination:** for eyes and skin and location of first aid supplies
**Emergency Procedures:** the procedures for handling non-routine tasks or emergency situations such as spills or fire
**Common Symptoms of Overexposure:** common symptoms of pesticide poisoning and ways poisoning can occur
**Exposure Hazards:** including both acute and chronic effects
**Environmental Concerns:** such as drift, runoff and wildlife hazards
**Laws and Regulations:** applicable laws and regulations, MSDS, Pesticide Safety Information Series, label requirements
**Employee Rights:** receive information on pesticides they may be exposed to, rights against discharge or other discrimination due to exercise of these rights
**Location of Documents:** Hazard Communication Program plans, pesticide use records, Pesticide Safety Information Series leaflets, MSDS and training records

[1]Insert the specific Brand Name of the rotenone product.

# Appendix B

Pesticide Training and Hazard Communication
CFT Legumine

Lake Davis
Pike Eradication Project

## Required Pesticide Training

- Health (immediate and long-term) effects
- Environmental concerns
- Pesticide label, MSDS, laws & regulations
- What to do in an emergency
- Personal Protective Equipment (PPE)
- Pesticide safety & handling procedures
- Rights as employee & where to find information on pesticides

## Pesticide Laws & Regulations

- Registered pesticides
  - Federal EPA registration
  - State registration
- 40 CFR – Environ Prot Agency (FIFRA)
- 21 CFR – Food & Drug Admin (residues)
- 29 CFR – Occ Safety & Health Admin (OSHA)
- Failure to comply w/ label - misdemeanor
  - Criminal (intentional or negligent)
    - Up to $5,000
    - 6 months in jail
  - Civil (unintentional) – up to $25,000
  - Administrative remedies – license revoked

## Section I
## Potential Acute & Chronic Effects

- Rotenone is a botanical compound
- Found in roots in the bean family
- Used for centuries for gathering fish
- Mode of action – interference with cellular respiration in mitochondria
- Cell → Organ → Organism

## Mammalian Toxicity & Signal Words (acute lethal oral dose)

| $LD_{50}$ (mg/kg) | Toxicity Rating | Toxicity Category | Signal Word |
|---|---|---|---|
| $\leq$ 50 | I | Highly toxic | Danger Poison |
| 51 – 500 | II | Moderately toxic | Warning |
| 501 – 2,000 | III | Slightly toxic | Caution |
| > 2,000 | IV | Practically nontoxic | Caution |

## Rotenone Acute Toxicity Profile

| Toxicity Value | Toxicity Category | Signal Word |
|---|---|---|
| Oral $LD_{50}$ rat 102 mg/kg (M)<br>320 mg/kg (F) | Category II<br>Category II | Warning<br>Warning |
| Dermal $LD_{50}$ rabbit < 5,000 mg/kg | Category III | Caution |
| Dermal Irritation rabbit | Category IV | Caution |
| Eye Irritation rabbit | Category IV | Caution |
| Inhalation rat $LD_{50}$ >0.060 mg/L (M)<br>>0.060 mg/L (F) | Category I<br>Category I | Danger/Poison<br>Danger/Poison |

## Rotenone Chronic Toxicity Profile

| Study | Effect |
|---|---|
| Oral 90-d (dog) NOEL = 0.4 mg/kg/d | Decreased growth |
| Teratogenicity (rat) & (mouse) | Not teratogenic |
| Reproduction (two-generation rat) | No effect |
| Chronic oncogenicity (rat) & (mouse) | Not carcinogenic |
| Gene mutation (bacteria) & (mouse bone marrow) | Not mutagenic |

## Summary of Rotenone Toxicity

- Rotenone moderately toxic (oral & inhalation)
- Little dermal & eye toxicity or irritation
- Not carcinogenic, teratogenic, or mutagenic
- No evidence rotenone causes Parkinson's Disease
  - Unrealistic exposure & conflict with other studies
  - Epidemiological
  - Occupational
- **risk** = $\int$ toxicity · exposure
  - Cannot control toxicity
  - Can control exposure

## CFT Legumine™ Hazard Summary

- Fatal if inhaled
- May be fatal if swallowed
- Substantial but temporary eye injury
- Causes skin irritation
- How to avoid:
  - Do not get in eyes, on skin or clothing
  - Do not breath spray mist (rotenone not volatile)

## Section II
## Pesticide Safety & PPE

- Working safely with pesticides
  - Read and follow label directions
  - Be careful with undiluted pesticides
  - Wear the right type of protection
- CFT Legumine™ Label
  - Signal word – Warning (Category II)
  - Precautionary statements
  - First aid instructions
  - Use instructions (rate, mixing & loading)

## Major Routes of Pesticide Exposure

- Inhalation – breathing volatile product vapors & dust
- Ingestion – swallowing treated water or product
- Absorption – treated water or product through the skin
- Eyes – splashing of treated water or product

## CFT Legumine Label

**PRECAUTIONARY STATEMENTS**
**HAZARDS TO HUMANS AND DOMESTIC ANIMALS**
**WARNING**

May be fatal if inhaled or swallowed. Causes moderate eye irritation. Harmful if absorbed through skin. Do not breathe spray mist. Do not get in eyes, on skin, or on clothing. Wear goggles or safety glasses.

When handling undiluted product, wear either a respirator with an organic-vapor-removing cartridge with a prefilter approved for pesticides (MSHA/NIOSH approval number prefix TC-23C), or a canister approved for pesticides (MSHA/NIOSH approval number prefix 14G), or a NIOSH approved respirator with an organic vapor (OV) cartridge or canister with any R, P, or HE prefilter.

Wash thoroughly with soap and water after handling and before eating, drinking, or using tobacco. Remove contaminated clothing and wash before reuse. Prolonged or frequently repeated skin contact may cause allergic reactions in some individuals.

**ENVIRONMENTAL HAZARDS**

This pesticide is extremely toxic to fish. Fish kills are expected at recommended rates. Consult your State Fish and Game Agency before applying this product to public waters to determine if a permit is needed for such an application. Do not contaminate untreated water when disposing of equipment washwaters.

**CHEMICAL AND PHYSICAL HAZARDS**

FLAMMABLE. KEEP AWAY FROM HEAT AND OPEN FLAME. FLASH POINT MINIMUM 45°F (7°C).

**For information on this pesticide product (including health concerns, medical emergencies, or pesticide incidents), call the National Pesticide Information Center at 1-800-858-7378.**

**STORAGE AND DISPOSAL**

Do not contaminate water, food or feed by storage or disposal.

**STORAGE:** Store only in original containers, in a dry place inaccessible to children and pets. This product will not solidify nor show any separation at temperatures down to 40°F and is stable for a minimum of one year when stored in sealed drums at 70°F.

**PESTICIDE DISPOSAL:** Pesticide wastes are acutely hazardous. Improper disposal of excess pesticide, spray mixture, or rinsate is a violation of Federal law. If these wastes cannot be disposed of by use according to label instructions, contact your state pesticide or Environmental Control Agency, or the Hazardous Waste representative at the nearest EPA Regional Office for guidance.

**CONTAINER DISPOSAL:** Triple rinse or equivalent. Then offer for recycling or reconditioning, or puncture and dispose of in a sanitary landfill, or by other procedures approved by state and local authorities.

**DIRECTIONS FOR USE**

**It is a violation of Federal law to use this product in a manner inconsistent with its labeling.**

**CFT Legumine is registered for use by or under permit from, and after consultation with State and Federal Fish and Wildlife Agencies.**

**GENERAL INFORMATION**

**This product is a specially formulated product containing rotenone to be used in fisheries management for the eradication of fish from lakes, ponds, reservoirs and streams.**

August 9, 2007 Master 2

## CFT Legumine Label

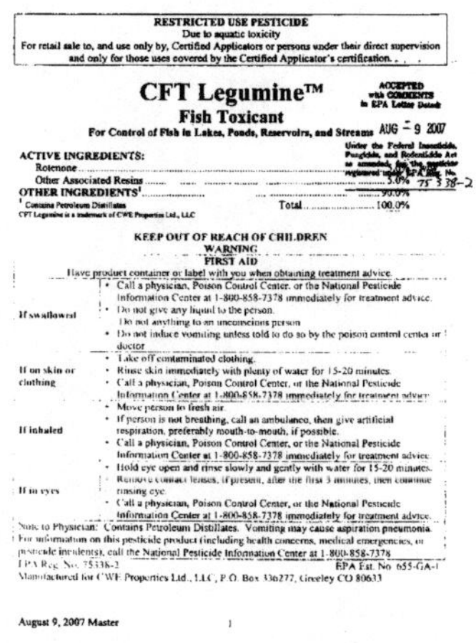

**RESTRICTED USE PESTICIDE**
Due to aquatic toxicity
For retail sale to, and use only by, Certified Applicators or persons under their direct supervision and only for those uses covered by the Certified Applicator's certification.

**CFT Legumine™**
**Fish Toxicant**
**For Control of Fish in Lakes, Ponds, Reservoirs, and Streams**

ACCEPTED with COMMENTS In EPA Letter Dated AUG 9 2007 Under the Federal Insecticide, Fungicide, and Rodenticide Act as amended, for the pesticide registered under EPA Reg. No. 75338-2

**ACTIVE INGREDIENTS:**
Rotenone
Other Associated Resins ........ 5.0%
**OTHER INGREDIENTS**[1] ........ 90.0%
Total ........ 100.0%

[1] Contains Petroleum Distillates
CFT Legumine is a trademark of CWE Properties Ltd., LLC

**KEEP OUT OF REACH OF CHILDREN**
**WARNING**
**FIRST AID**

Have product container or label with you when obtaining treatment advice.

| | |
|---|---|
| If swallowed | • Call a physician, Poison Control Center, or the National Pesticide Information Center at 1-800-858-7378 immediately for treatment advice.<br>• Do not give any liquid to the person.<br>Do not anything to an unconscious person<br>• Do not induce vomiting unless told to do so by the poison control center or doctor |
| If on skin or clothing | • Take off contaminated clothing.<br>• Rinse skin immediately with plenty of water for 15-20 minutes.<br>• Call a physician, Poison Control Center, or the National Pesticide Information Center at 1-800-858-7378 immediately for treatment advice. |
| If inhaled | • Move person to fresh air.<br>• If person is not breathing, call an ambulance, then give artificial respiration, preferably mouth-to-mouth, if possible.<br>• Call a physician, Poison Control Center, or the National Pesticide Information Center at 1-800-858-7378 immediately for treatment advice. |
| If in eyes | • Hold eye open and rinse slowly and gently with water for 15-20 minutes.<br>• Remove contact lenses, if present, after the first 5 minutes, then continue rinsing eye.<br>• Call a physician, Poison Control Center, or the National Pesticide Information Center at 1-800-858-7378 immediately for treatment advice. |

Note to Physician: Contains Petroleum Distillates. Vomiting may cause aspiration pneumonia.
For information on this pesticide product (including health concerns, medical emergencies, or pesticide incidents), call the National Pesticide Information Center at 1-800-858-7378

EPA Reg. No. 75338-2 EPA Est. No. 655-GA-1
Manufactured for CWE Properties Ltd., LLC, P.O. Box 336277, Greeley CO 80633

August 9, 2007 Master 1

## Wearing Right Kind of Protection

- Goggles
  - Required when mixing, loading & applying
  - Safety glasses, face shield, full-face mask
- Chemically resistant gloves
  - Required when mixing, loading & applying
  - Never fabric-lined
- Coveralls or Tyvek suit (potential heat illness)
  - Required when using Class I or II pesticides
  - Chemically repellent
  - New suit each day
  - Heat illness reduction (slits in armpit & inseam for ventilation)
- Full-face respirator (handling undiluted product)
  - Required on label
  - Rotenone & solvent are relatively nonvolatile

## Sharp-Dressed Applicator (Tyvek suit, gloves & goggles)

## Pesticides on Hands

- Keeping pesticides off of hands
- Hands
  - Eyes when rubbing
  - Mouth when touching food
- Frequently wash hands
  - Eating
  - Drinking
  - Smoking

## Section III
## Environmental Concerns

- Rotenone toxic to fish – fish kills expected
- Cleaning of equipment & disposal of wash waters in treatment area only
- Rotenone not toxic to wildlife @ use rate
- Low BCF and vapor pressure
- Hydrolysis half-lives in days
- Photolysis half-lives in hours
- Abiotic & biotic degradation

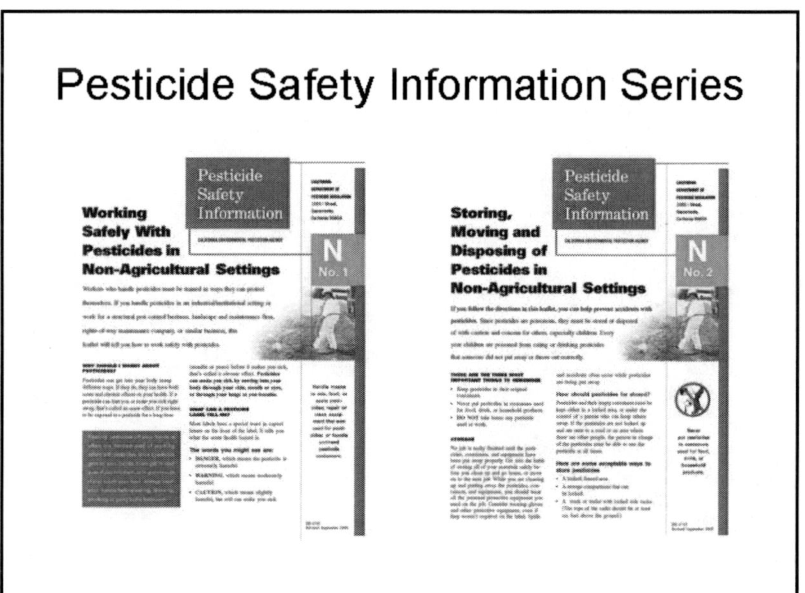

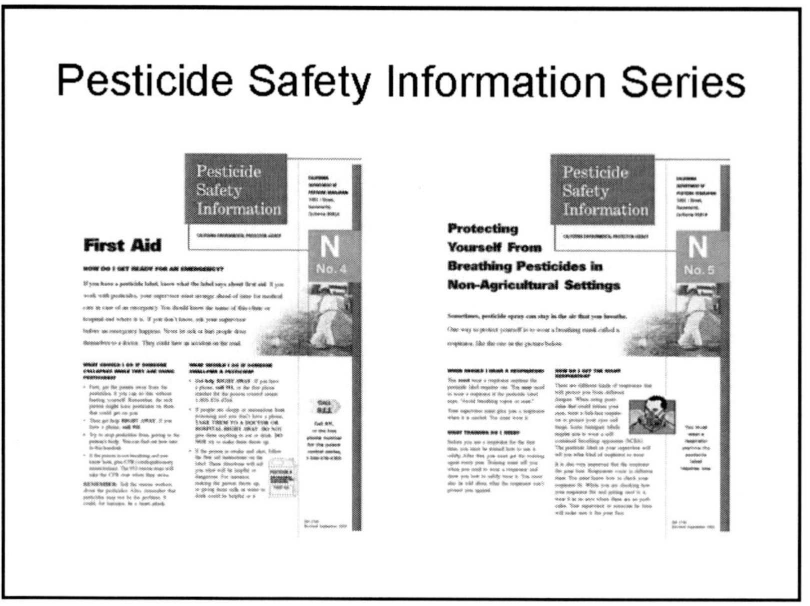
Pesticide Safety Information Series
Pesticide Safety Information
First Aid
N
No. 4
Pesticide Safety Information
Protecting Yourself From Breathing Pesticides in Non-Agricultural Settings
N
No. 5

Pesticide Safety Information Series
Pesticide Safety Information
Washing Pesticide Work Clothing
N
No. 7
Pesticide Safety Information
Safety Rules for Pesticide Handlers in Non-Agricultural Settings
N
No. 8

# CFT Legumine MSDS

CWE Properties Ltd., LLC – P.O. Box 336277 – Greeley, CO 80633

**CFT Legumine™** EPA Reg. No. 75338-2

**Material Safety Data Sheet**

**SECTION 1: CHEMICAL PRODUCT AND COMPANY IDENTIFICATION**

**PRODUCT/CHEMICAL NAME:** CFT Legumine™

**Emergency Contact:** 1-800-858-7378 (National Pesticide Information Center)

**Transportation Emergency Contact:** 1-800-858-7378 (National Pesticide Information Center

**Manufactured for: CWE Properties Ltd., LLC**
P.O. Box 336277
Greeley, CO 80633

**SECTION 2: HAZARDS IDENTIFICATION SUMMARY**

KEEP OUT OF REACH OF CHILDREN—WARNING – May be fatal if inhaled. May be fatal if swallowed. Causes substantial, but temporary, eye injury. Causes skin irritation. Do not breathe spray mist. Do not get in eyes, on skin, or on clothing. Wear goggles or safety glasses. This product is an orange, viscous liquid with slight petroleum odor.

**SECTION 3: COMPOSITION / INFORMATION ON INGREDIENTS**

| Chemical Ingredients: | Percentage By Weight | CAS No. | TLV (Units) |
|---|---|---|---|
| Rotenone | 5.00 | 83-79-4 | 5 mg/m$^3$ |
| Other Associated Resins | 5.00 | | |
| Inert Ingredients, including N-Methylpyrrolidone | 90.00 | 872-50-4 | not listed |

**SECTION 4: FIRST AID MEASURES**

**IF SWALLOWED:** Call a physician, Poison Control Center, or the National Pesticide Information Center at 1-800-858-7378 immediately for treatment advice. Do not induce vomiting unless told to do so by the Poison Control Center or physician. Do not give any liquid to the person. Do not give anything by mouth to an unconscious or convulsing person.

**IF INHALED:** Remove victim to fresh air. If not breathing, give artificial respiration, preferably by mouth-to-mouth. Call a physician, Poison Control Center, or the National Pesticide Information

Emergency Telephone Number: 1-800-858-7378

Revision Date: July 12, 2007

Page 1 of 5

# Section IV Rights & Access to Label & MSDS

- Lake Davis Hazard Communication Book
  - CFT Legumine™ label
  - CFT Legumine™ Material Safety Data Sheet
  - Pesticide Safety Information Series
  - Summary of Public Health Studies
  - Summary of Environmental Fate & Effects
- Pesticide Hazard Communication Program (PIU)
  - Pesticide use records of DFG
  - Pesticide Safety Information Series
  - Material Safety Data Sheets & Labels
  - Training records

# Pesticide Safety Training Record

CALIFORNIA DEPARTMENT OF FISH AND GAME

PESTICIDE SAFETY TRAINING RECORD

This is to certify that ______________________ has received __________ hours of pesticide safety training.

Employee Signature ______________________ Employee Title ______________________

Employee Work Location ______________ Trainer ______________ Date ______________

TRAINING REQUIREMENTS

**SAFETY PROCEDURES**: personal protective equipment, engineering controls, and equipment. Heat-related illnesses.
**PESTICIDE LABELS**: signal words, precautionary statements, first aid instructions, mixing and application instructions.
**PESTICIDE HANDLING PROCEDURES**: container handling, mixing and application equipment, triple rinse containers.
**FIRST AID AND DECONTAMINATION**: for eyes and skin and location of first aid supplies.
**EMERGENCY PROCEDURES**: the procedures for handling non-routine tasks or emergency situations such as spills or fire.

**COMMON SYMPTOMS OF OVEREXPOSURE**: common symptoms of pesticide poisoning and ways poisoning can occur.
**EXPOSURE HAZARDS**: including both acute and chronic effects.
**ENVIRONMENTAL CONCERNS**: such as drift, runoff and wildlife hazards.
**LAWS AND REGULATIONS**: applicable laws and regulations, Material Safety Data Sheets, Pesticide Safety Information Series leaflets, label requirements.
**EMPLOYEE RIGHTS**: to receive information on pesticides they may be exposed to, rights against discharge or other discrimination due to exercise of these rights.
**LOCATION OF DOCUMENTS**: Hazard Communication Program plans, pesticide use records, Pesticide Safety Information Series leaflets, MSDS and training records.

PESTICIDES

- ☐ **2,4-D** (Weedar® 64, Weedone® LV4)
- ☐ **Aminopyralid** (Milestone®)
- ☐ **Chlorsulfuron** (Telar®)
- ☐ **Clopyralid** (Transline®)
- ☐ **Diquat** (Reward®)
- ☐ **Diuron** (Direx®, Karmex®)
- ☐ **Fluazifop p-butyl** (Fusilade®)
- ☐ **Glyphosate** (AquaMaster®, Roundup® Pro)
- ☐ **Imazapyr** (Stalker®, Habitat®)
- ☐ **Oxyfluorfen** (Goal®)
- ☐ **Rotenone** (CFT Legumine™, Nusyn-Noxfish®)
- ☐ **Simazine** (Princep®)
- ☐ **Triclopyr** (Garlon®, Pathfinder®, Renovate® 3)
- ☐ **Trifluralin** (Treflan®)

Rev. 8/07 FG1075

# Rotenone Storage, Transportation, and Spill Containment
# SOP:4.0

**PROCEDURE TITLE:** Rotenone Storage, Transportation, and Spill Containment

**APPLICABILITY:** Application of rotenone in waters of the United States

**PURPOSE:**

1. Mitigate for environmental contamination resulting from rotenone spillage during storage and transportation
2. Provide a protocol for safe and effective storage, transportation, and spill prevention and containment

**LOCATION ON LABEL:** "Storage and Disposal"

**PROCEDURE:**

I. Storage

A. Original and Service Containers

Store rotenone in original containers or in approved service containers, in a dry place and inaccessible to children and pets. Service containers can be any secure and operational storage container except those of a type commonly used for food, drink or household products. Service containers provide (1) the name and address of the person or firm responsible for the container, (2) the identity of the pesticide in the container, and (3) the signal word that appears on the label of the original container (i.e., "Danger," "Warning," or "Caution"). The labeling on the original container and any service container must be kept intact throughout the use of the container.

B. Guidelines for Long-term Storage Facilities

Pesticide storage areas should be in a separate room from office and residential spaces, water supply sources, and food or feed storage areas. Long-term storage of pesticides and pesticide containers should be in a fire-resistant structure with good ventilation and a sealed, concrete floor that slopes toward drainage and secondary containment. Pesticide storage areas should have separate entries if possible. Pesticide storage areas should have security and access control provisions including a locked door and locked windows (or no windows) to prohibit access. If liquid products are to be stored, the storage area should have a containment system capable of containing at least 25% of the stored liquid volume. Typical containment systems include a bermed floor or a sloped floor with a sump. The building temperature should be kept lower than 95°F (35°C) and above pesticide freezing points. All electrical fixtures and appliances in the storage area should be non-sparking units approved for use in facilities storing flammable and combustible liquids, if applicable. Weatherproof signs stating "Danger—Pesticides—Keep Out" or similar warning should be posted on each door and any window. Post the name, address, and phone number of a contact person at the primary entrance to the storage area. Signs visible from any direction of probable approach shall be posted around all storage areas where containers that hold, or have held, pesticides required to be labeled with the signal words "warning" or "danger" (Category I or II pesticides) are stored. Each sign shall be of such size that it is readable at a distance of 25 (7.6 m) feet and be substantially as follows:

DANGER
POISON/PESTICIDE STORAGE AREA
ALL UNAUTHORIZED PERSONS KEEP OUT
KEEP DOOR LOCKED WHEN NOT IN USE

The notice shall be repeated in an appropriate language other than English when it may reasonably be anticipated that persons who do not understand the English language will come to the enclosure.

C. Guidelines for Project Site

Acceptable storage enclosures may be a closed vehicle, closed trailer, a building, room, or fenced area with a fence at least six feet high, a foot locker or other container that can be locked, and trucks or trailers that have solid sideracks and secured tailgates at least six feet above ground, ramp, or platform level. Metal containers with screwtype bungs and/or secured and locked valves and sealed five-gallon containers are also acceptable storage enclosures. Because storage requirements may vary, consult your State pesticide regulation office to ensure compliance with local rules.

II. Transportation

A. Federal Regulation

The legal requirements regarding hazardous material transport, required training, placarding and special licenses are contained in Title 49 of the Code of Federal Regulations (49 CFR). The MSDS for liquid formulations of rotenone state that they are class 6.1 (poisonous materials) and marine pollutants. Depending upon the formulation, a liquid rotenone product may also be classed as a flammable liquid. Powdered rotenone is class 6.1 and also considered a marine pollutant. Due to the considerable environmental risks associated with the transport of rotenone formulations, it is crucial that commercial carriers must have the required training and credentials required by 49 CFR.

B. State Exemption

The federal codes regulating transport of hazardous materials exempts state agencies from many of the requirements regarding hazardous materials training, license endorsements and placarding. However, given the need to protect the public and the environment when transporting hazardous materials, it is recommended that public agencies comply with the federal codes to the fullest extent possible.

Commercial hazardous material carriers should be contracted when undertaking large projects if government carriers are not fully trained and not appropriately credentialed.

C. Restrictions

Pesticides shall not be transported in the same compartment with persons, food or feed. Pesticide containers shall be secured to vehicles during transportation in a manner that will prevent spillage onto the vehicle or off the vehicle. Paper, cardboard, and similar containers shall be covered (i.e., plastic tarp) when necessary to protect them from moisture. Carry shipping papers in your vehicle including emergency response phone numbers and MSDSs must be present in the vehicle when transporting rotenone. Prior to transport, the applicator or vehicle driver must ensure that pesticide containers are properly labeled.

D. Government Agency Documentation

It is recommended that federal, state, or local government workers transporting rotenone keep a document in the transportation vehicle stating their compliance with federal Hazardous Materials Regulations (HMR):

In accordance with § 49 CFR 171.1 (d)(5), which states:

(d) *Functions not subject to the requirements of the HMR.*
The following are examples of activities to which the HMR do not apply:

...(5) Transportation of a hazardous material in a motor vehicle, aircraft, or vessel operated by a federal, state or local government employee solely for noncommercial Federal, state, or local government purposes.

[this agency] is exempt from the HMR (§ 49 CFR 171 through 180) as it applies to the Packaging, Pre-transportation, and Transportation functions of hazardous materials.

Please contact the [agency] Safety Office at [telephone or other contact] if there are any concerns regarding this exemption.

III. Spill Prevention and Containment

A. Spill Prevention

Off-site spills may be associated with improper storage or accidents during handling and transport. Generally, all spills must be reported to the state spill response unit and other units as appropriate. Small spills may be contained and the collected material disposed of according to the product label. If these wastes cannot be disposed of by use according to label instructions contact your state pesticide or environmental control agency, or the hazardous waste representative at the nearest EPA Regional Office for guidance.

It is recommended that applicators create a Spill Contingency Plan (Section IV). Spill contingency plans can help avoid or minimize on-site spills. The complexity of the spill contingency plan will depend upon the size and detail of the rotenone treatment. Large projects will require detailed plans that consider project security and avoidance of catastrophic spills. The spill contingency plan should detail where the material will be stored on site while awaiting application.

B. Spill Containment

The storage of rotenone materials at the project location may be in a location that is graded to allow drainage to the project water body in case of an accidental spill. Containers of rotenone powder and liquid may be set on a plastic barrier, concrete ramp, or other impermeable surface sloped toward the project water body. A small spill of rotenone can then be rinsed into the treated water. The designated storage area on-site should be bermed and should be large enough to contain all the stored material. This will allow recovery of all the material. The berm may be constructed of straw/hay bales or other suitable material and should be lined with heavy duty plastic fabric. Portable bilge pumps, hoses, buckets, drums, absorbent clay and absorbent pads and other recovery equipment as well as personal protective equipment should be maintained in an adjacent area readily available in case of a spill. Each person who controls the use of any property or premises that holds, or has held rotenone, is responsible for all containers or equipment on the property. Unless all such containers are under personal control so as to avoid contact by unauthorized persons, make arrangements to (1) provide a person responsible to maintain such control over the containers at all times or (2) store all such containers in a locked enclosure, or in the case of liquid pesticides in a container larger than 55 gallons (208 L) in capacity, the container shall have a locked closure. Either shall be adequate to prevent unauthorized persons from gaining access to any of the material.

C. Spill Management

In the event of a spill, it is extremely important that the spilled material be contained. If a ground spill occurs, immediately control the spill at its source and contain or channelize the spilled material into a containment area with shovels and other hand tools. Once the material is contained or diked into pools, the applicator should attempt to recover the material by using absorbent materials such as clay, soil, sawdust or straw to absorb pooled liquids or collection by pump or sponge. Recovered material can be applied to the treatment area according to label instructions and other local, state and federal regulations.

IV. Spill Contingency Plan

The Spill Contingency Plan should contain the following elements:

- Inventory of materials to be used in the treatment (products, location, and amount).
- Description of storage areas (size, construction details, and security measures).
- Description of staging areas, mixing areas, treatment areas, and deactivation areas.

- Precautions—specific to the site, locale, and treatment.
- Chain of Command—details the flow of information and responsibility for all facets of the treatment.
- Contact information for all downstream water users who would be impacted in the event of a major spill.
- Contact information of all entities that must be contacted in the event of a reportable spill. This list is to be provided to all project personnel and should be with them at all times.
- Specific spill containment and recovery procedures.
- Indicate mode of communication and all referenced areas on a map.

## V. Additional Information

University of Nevada, Reno, Cooperative Extension. Safe and Legal Transportation of Pesticides, Special Publication SP-01-09. Available: www.unce.unr.edu/publications/files/ag/2001/sp0109.pdf.

University of Nebraska, Lincoln. Safe Transport, Storage, and Disposal of Pesticides. Extension Publication ED2507. Available: www.ianrpubs.unl.edu/epublic/live/ec2507/build/ec2507.pdf.

University of California, Statewide Integrated Pest Management Program. 2000. The Safe and Effective Use of Pesticides, 2nd edition. Agricultural and Natural Resources Publication 3324.

# Determining Treatment Rates and Strategies SOP:5.0

**PROCEDURE TITLE:** Determining Treatment Rates and Strategies

**APPLICABILITY:** Application of rotenone in waters of the United States

**PURPOSE:**

1. Provide strategies for eliminating undertreatment and overtreatment of target species
2. Provide guidance on conducting bioassays and designing treatments using effective pest management techniques

**LOCATION ON LABEL:** "Determining Treatment Rate" under the heading "Directions for Use"

**PROCEDURE:**

I. Coverage of Water Treated Using Labeled Rates

The actual treatment rate and rotenone concentration needed to kill a target species varies widely, depending on the type of water, environmental factors including pH, temperature, depth, and turbidity, and sensitivity of target species. Tables SOP 5.1 (Table 5.2 for metric) and SOP 5.3 (Table 5.4 for metric) are general guides for proper rates and rotenone concentrations for eradication of target species (i.e., complete kills of target species) and provide estimates of expected coverage in either standing or flowing waters. The Certified Applicator (or someone acting under their direct supervision) must conduct bioassays using site water (or water of similar quality) and target species (or surrogate species of similar sensitivity) to refine the treatment rate within the maximum limit allowed. Treatment concentrations above the label maximum of 200 ppb active rotenone are a violation of federal law under FIFRA.

Table SOP 5.1. Recommended rotenone treatment concentrations and number of acre-feet (AF) standing water covered by one gallon or one pound of (5% a.i.) product. Adjust amount of product according to the actual rotenone content on Ingredient Statement on label.

| Type of Use | Parts per Million (ppm) | | AF | AF |
|---|---|---|---|---|
| | Product (5% a.i.) | Active Rotenone | per Gallon Liquid | per Pound Powder |
| Normal | 0.5–1.0 | 0.025–0.05 | 6.0 to 3.0 | 0.74 to 0.37 |
| Tolerant Species | 1.0–3.0 | 0.05–0.15 | 3.0 to 1.0 | 0.37 to 0.123 |
| Tolerant Species in Organic Ponds | 2.0 – 4.0 | 0.1 – 0.2 | 1.5 to 0.75 | 0.185 to 0.093 |

TABLE SOP 5.2. Recommended rotenone treatment concentrations and number of cubic meters ($m^3$) standing water covered by one liter or kilogram of (5% a.i.) product. Adjust amount of product according to the actual rotenone content on Ingredient Statement on label.

| Type of Use | Parts per Million (ppm) | | $m^3$ | $m^3$ |
|---|---|---|---|---|
| | Product (5% a.i.) | Active Rotenone | per Liter Liquid | per Kilogram Powder |
| Normal | 0.5–1.0 | 0.025–0.05 | 2000 to 1000 | 2000 to 1000 |
| Tolerant Species | 1.0–3.0 | 0.05–0.15 | 1000 to 333 | 1000 to 333 |
| Tolerant Species in Organic Ponds | 2.0–4.0 | 0.1–0.2 | 500 to 250 | 500 to 250 |

TABLE SOP 5.3. Recommended rotenone treatment concentrations and number of cubic feet per second ($ft^3/s$) flowing water treated for 4- and 8-hour (hr) periods with one gallon of (5% a.i.) product. Adjust amount of product according to the actual rotenone content on Ingredient Statement on label.

| Type of Use | Parts per Million (ppm) | | $ft^3/s$ | $ft^3/s$ |
|---|---|---|---|---|
| | Product (5% a.i.) | Active Rotenone | per Gallon (4-hr) | per Gallon (8-hr) |
| Normal | 0.5–1.0 | 0.025–0.05 | 18.4 to 9.2 | 9.2 to 4.6 |
| Tolerant Species | 1.0–3.0 | 0.05–0.15 | 9.2 to 3.1 | 4.6 to 1.6 |
| Tolerant Species in Organic Waters | 2.0–4.0 | 0.1–0.2 | 4.6 to 2.3 | 2.3 to 1.2 |

TABLE SOP 5.4. Recommended rotenone treatment concentrations and number of cubic meters per second ($m^3/s$) flowing water treated for 4- and 8-hour (hr) periods with one liter of (5% a.i.) product. Adjust amount of product according to the actual rotenone content on Ingredient Statement on label.

| Type of Use | Parts per MIllion (ppm) | | $m^3/s$ | $m^3/s$ |
|---|---|---|---|---|
| | Product (5% a.i.) | Active Rotenone | per Liter (4-hr) | per Liter (4-hr) |
| Normal | 0.5–1.0 | 0.025–0.05 | 0.138 to 0.069 | 0.069 to 0.034 |
| Tolerant Species | 1.0–3.0 | 0.05–0.15 | 0.069 to 0.024 | 0.034 to 0.013 |
| Tolerant Species in Organic Waters | 2.0–4.0 | 0.1–0.2 | 0.034 to 0.018 | 0.0180 to 0.008 |

## II. Using Bioassay to Determine Site-Specific Treatment Rate

Determine treatment rate, within maximum limits allowed (Tables 5.1 through 5.4), by conducting a bioassay with the target species (or surrogate species of similar sensitivity) in the site water (or water quality equivalent i.e., pH, temperature, turbidity). Ideally, this affirmation of potency would be conducted in the laboratory under controlled conditions, but may be completed in the field immediately prior to treatment.

For example, a bioassay can be completed using 10-gallon (37.9-L) aquaria or 5-gallon (18.9-L) plastic buckets containing 20 or 10 L of water each, respectively. The following active rotenone concentrations would be tested: 0.0 (control), 12.5, 25, 50, 100, and 200 ppb. To test these active ingredient concentrations, make a 50 ppm active ingredient (1,000 ppm product) Rotenone Stock Solution by diluting 1 ml of liquid concentrate or 1 g of powder to 1 L of water and mixing thoroughly. Add the amounts indicated in Table 5.5 to 10 or 20 L of test volume.

Add 3 to 10 fish (not to exceed 1g fish per liter of water) and monitor survival at 30, 60, 120, 240 and 480 minutes (0.5, 1, 2, 4, and 8 hours). Keep test solutions near ambient water temperature with the use of refrigeration or by partially submerging aquaria or buckets in water body if necessary. If the fish loading is exceeded for the size of the container, then use a larger container containing proportionately more water and stock solution. Normally, aeration of test water during the bioassay is not recommended as this may hasten the breakdown on rotenone.

The lowest active rotenone concentration that produces 100% mortality of test fish within the treatment duration is the Minimum Effective Dose (MED). For standing water treatments, use mortality estimate for 8 hours exposure as MED. For flowing water, use the expected treatment interval. Depending on the pH, turbidity, temperature, sunlight intensity, and depth, treat at a concentration at least twice the MED to ensure complete kills of target fish. Alkaline pH, high temperature, sunlight intensity, turbidity, and dispersion into deep water all affect the dissipation of rotenone. Ensure that maximum rates listed in Tables SOP 5.1 through SOP 5.4 are not exceeded.

Make sure that measuring devices, aquaria, and other equipment coming in contact with rotenone are cleaned prior to use. Following use, clean all equipment with soap and water and let dry in sun light prior to reuse. Deactivation with a 1% potassium permanganate solution followed by complete rinsing will assure total elimination of rotenone residues.

Wear required safety gear according to label and MSDS when handling rotenone product, dilute rotenone solutions, and potassium permanganate.

TABLE SOP 5.5. Amount (in ml) of 50 ppm Rotenone Stock Solution (1 ml or 1 g product to 1 liter water) needed to achieve various concentrations of active rotenone in solutions containing 10 and 20 liters (control solution receives no Rotenone Stock Solution).

| Test Volume | Active Rotenone 0.0125 ppm | Active Rotenone 0.025 ppm | Active Rotenone 0.050 ppm | Active Rotenone 0.10 ppm | Active Rotenone 0.20 ppm |
|---|---|---|---|---|---|
| 10 Liters | 2.5 | 5 | 10 | 20 | 40 |
| 20 Liters | 5 | 10 | 20 | 40 | 80 |

Dispose of test fish properly by burying on-site in the ground away from the treated water body or through normal laboratory disposal procedures.

## III. Strategies for Removal of Target Fish

Exposing all target fish to sufficient rotenone concentration and duration should result in a complete kill of the target fish species. All habitats capable of supporting target fish within the area should be treated unless there is conclusive evidence that target fish are absent or that the area cannot serve as a safe haven for target fish during the treatment. The consequences of not treating an area in question within the eradication zone far outweigh those of treating the area in question. Physical removal techniques including netting, electrofishing, explosives and traps all have efficiency biases, and negative results are not conclusive evidence of absence of target fish.

For treatment of streams, habitat includes headwater lakes and all tributaries, seeps and springs that are or can be hydrologically connected to the area scheduled for eradication of target fish. Avoid the flow of untreated water into the treatment area during the treatment; ideally the entire area concurrently contains lethal levels of rotenone. Normally, place drip stations every one to two hours travel time (or as determined by bioassay) apart on flowing reaches (see SOP 11) and spray seeps, backwater areas and springs with dilute rotenone (see SOP 12) or treat with rotenone powder/gelatin/sand mixture (see SOP 13).

Complex habitats (seeps, springs, backwaters, weedy areas) increase the potential that rotenone will not reach lethal concentrations in all locations with the result that some fish may not be killed. For this reason, it is recommended that multiple treatments be employed. The treatments should be adequately spaced in time to allow eggs time to hatch and develop so they become vulnerable to a second treatment. Spacing treatments also allows time for physical conditions (vegetation, flows, water levels) to change which will change exposure conditions for the fish. Generally, multiple applications of rotenone over several years are required for complete removal of target species from a stream or river.

For treatment of lakes, habitat includes the entire water body and upstream tributaries. Normally, treat an entire lake within 48 hours to insure that rotenone does not degrade during application. Dispersion of rotenone horizontally and vertically in lakes is required for complete fish kills. Impediments to dispersion of rotenone include backwater and weedy areas that receive little or no water movement, and deep lakes where there is little water exchange through a thermocline. Spray backwater and weedy areas with dilute rotenone and pumping rotenone to depth may be required to get it through the thermocline. Identify springs and seeps that may provide sanctuary to target species and treat with liquid rotenone or the rotenone powder/gelatin/sand mixture (see SOP 13). Generally, one application of rotenone is sufficient for complete removal of target species from a lake or pond.

## IV. Sentinel Fish to Monitor Effectiveness of Rotenone

Normally, place cages containing fish upstream of all drip cans in streams and at various locations and depths in lakes to monitor the effectiveness of the treatment. Target areas where poor water movement or premature breakdown of rotenone is expected such as waterfalls in streams and weedy areas of lakes. Careful monitoring of sentinel fish during or following the application is used to adjust treatment rate and application site and method.

## V. Strategic Considerations for Complete Eradication of Target Fish Species

- Treat all known habitat within project area capable of supporting fish
- Avoid undertreatment and escapement
- Avoid overtreatment, excessive rotenone residues, and inadequate deactivation
- Use liquid rotenone in streams and lakes
- Use powdered rotenone only in lakes
- Use accurate and up-to-date volume, depth, and surface area measurements of lakes
- Use accurate and up-to-date flow, velocity, and length of streams
- Identify all barriers to fish and water movement
- Identify all types and densities of aquatic vegetation
- Identify all inlets/outlets and seeps/springs
- Identify physical and chemical characteristics of site water
- Use block nets where appropriate to prevent target fish escapement

# Determining Treatment Areas and Project Areas SOP:6.0

**PROCEDURE TITLE:** Determining Treatment Areas and Project Areas

**APPLICABILITY:** Application of rotenone in waters of the United States

**PURPOSE:** Provide criteria for determining the treatment and project areas to assure public safety and legal requirements are met

**LOCATION ON LABEL:** "Placarding of Treatment Areas" under the heading "Directions for Use"

**PROCEDURE:**

I. Determining Treatment Area

The Treatment Area is the water body where the application of rotenone is expected to produce a total kill of the target species. For stream treatments with multiple drip stations, the Treatment Area extends downstream from the lowermost drip station to a point where killing of fish is no longer desired; generally, equivalent to the distance between drip stations. Exceptions to this recommendation include (1) where the stream goes subsurface below the last drip station and the point where surface flow ceases becomes the lower margin of the Treatment Area; (2) where rotenone-free surface flow joins the treated water and results in a dilution that is judged non-toxic and the point of confluence becomes the lower margin of the Treatment Area; and (3) at the point where a deactivation station is positioned downstream of the lowermost drip station. Activities restricted to the Treatment Area include:

A. Wash or rinse all application equipment in the Treatment Area (see "Hazards to Humans and Domestic Animals" under the heading "Precautionary Statements").

B. Workers re-entering Treatment Areas where applications are >0.09 ppm (>90 ppb) active rotenone refer to "Placarding of Treatment Areas" and "Re-entering the Treatment Area" under "Directions for Use" and SOP 1.0 for information on when placards and protective PPE are no longer required.

C. Recreational access is not allowed within the Treatment Area during the application of rotenone (see "Placarding of Treatment Areas" under "Directions for Use").

D. For applications >0.04 ppm (>40 ppb) active rotenone in waters with drinking water intakes or hydrologic connections to wells, certain notifications and monitoring requirements apply (see "Drinking Water Monitoring" under "Directions for Use" and SOP 16.0).

## II. Determining Project Area

The Project Area includes the Treatment Area plus all other surface and groundwater areas in hydrologic connection that would be affected by rotenone application and deactivation. The area should also include any location where the public or project personnel will be exposed to drift from an aerial application of rotenone or where the rotenone is deactivated by potassium permanganate. It also includes adjoining areas that will be affected by the presence and activity of personnel associated with the project. The Project Area is not defined by the label but is a useful demarcation of the impacted area for which effects are analyzed during the EA. The restrictions in the Project Area should be the same as those in the Treatment Area (see "Placarding of Treatment Areas" under "Directions for Use" and SOP 1.0).

In addition:

A. The Certified Applicator must remain in Project Area during the rotenone application phase of the treatment.

B. If a hydrologic connection exists for domestic wells within Project Area, notification to users must be provided 7–14 days prior to application (see "Drinking Water Notification" and "Drinking Water Monitoring" under "Directions for Use" and SOP 16.0).

C. The Project Area includes the deactivation zone where potassium permanganate is used (see SOP 7.0).

D. The Project Area includes the landscape adjacent to the Treatment Area where biological effects caused by rotenone or deactivation can occur.

E. The Project Area includes the landscape adjacent to the Treatment Area where human activities related to the use of rotenone can occur. These include staging areas for equipment and personnel, storage of chemicals, camping/cooking areas, trash/refuse areas, location of pack stock, and helicopter landing spots.

F. For applications >0.04 ppm (40 ppb) active rotenone in waters with drinking water intakes or with hydrologic connections to wells, certain notification and monitoring requirements apply (see SOP 16.0). This applies to drinking water intakes and wells within the deactivation zone.

G. Aerial applications may produce drift which would extend beyond the shorelines or stream banks of the Treatment Areas. If drift is predictable, the affected landscape is part of the Project Area.

# Determining Need and Methods for Chemically Induced Deactivation SOP:7.0

**PROCEDURE TITLE:** Determining Need and Methods for Chemically Induced Deactivation

**APPLICABILITY:** Application of rotenone in waters of the United States

**PURPOSE:**
1. Mitigate for movement of rotenone and effects to non-target organisms beyond the Treatment Area (see SOP 6)
2. Provide guidance for determining if deactivation is necessary to protect non-target organisms beyond the Treatment Area and a protocol for successful deactivation

**LOCATION ON LABEL:** "Deactivation" under the heading "Directions for Use"

**PROCEDURE:**

I. Need for Deactivation

Rotenone treated water is deactivated to minimize exposure to non-target organisms beyond (e.g. downstream) the Treatment Area, unless determined "unnecessary" by the Certified Applicator. Guidance on the necessity of deactivation is explained below through examples where deactivation is considered "necessary" and "unnecessary." The Project Area includes the Treatment Area and the deactivation zone, if used (see SOP 6).

A. Necessity and Feasibility of Deactivation

If rotenone-treated discharge affects non-target species beyond the margins of the Treatment Area (see SOP 6.0 for definition), then it is necessary to deactivate the discharge unless:

1. If there is no discharge from the Treatment Area or the discharge goes dry in a distance shorter than 2 miles or 2 hours travel-time (maximum distance/time between drip stations recommended on label) from the lowest drip station or discharge, then it is unnecessary to deactivate the discharge. Examples include ponds or lakes with no discharge or a stream that goes dry (i.e., underground) a short distance (within 2 miles) from the Treatment Area.

2. If there are physical limitations that prohibit the operation of deactivation equipment, then it is unfeasible to deactivate the discharge. Examples include a stream where treated water flows into a canyon or chasm and access to the stream is difficult or safety is an issue. In such situations, the Certified Applicator through bioassay or analytical testing assures that the discharge is no longer toxic at 30 minutes travel time downstream of where the stream emerges at an accessible location and deactivation could be accomplished.

3. If dilution from untreated waters below the Treatment Area renders the discharge with undetectable concentrations of rotenone (defined as active rotenone < 2 ppb; see SOP 16) then it is unnecessary to deactivate the discharge. For example, a stream flowing at 3 $ft^3/s$ and treated at 25 ppb active rotenone flows into a stream flowing at 40 $ft^3/s$ and not treated with rotenone, yields a dilution fraction of 0.075 and an expected active rotenone concentration of 1.9 ppb (0.075 x 25 ppb = 1.875 ppb active rotenone), and it is not necessary to deactivate.

## II. Methods for Chemically-Induced Deactivation

Potassium permanganate ($KMnO_4$) is the only chemical allowed on the label for deactivating rotenone, applied in granular form or as a 2.5% solution (1 pound $KMnO_4$ to 5 gallons (25 g/L) water). A 2.5% solution assures that $KMnO_4$ remains in solution at most water temperatures. The solubility in distilled water is 1.8 pounds $KMnO_4$/5 gallons (43 g/L) water at 10°C and 2.7 pounds $KMnO_4$/5 gallons (65 g/L) water at 20°C.

Deactivation activity increases directly with increase in temperature, and applicators should take in account that chemical reactions slow by 50% for each 10°C reduction in temperature and increase by 2-fold for each 10°C increase (Q10 rule). This is important in determining the stream reach and deactivation lag times at very low temperatures. For example, twice the contact time will be required at a temperature of 5°C than at 15°C with the same level of $KMnO_4$.

$KMnO_4$ is toxic to fish (see SOP 14) at relatively low concentrations and is more toxic in alkaline water than soft water (Marking and Bills 1975). If $KMnO_4$ concentrations are in balance with rotenone concentrations, then toxic levels of $KMnO_4$ are reduced through the oxidation of organic components and rotenone.

### A. Application Rates

1. Deactivation of rotenone is a time-dependent reaction; the time the two chemicals are in contact dictates the effectiveness or degree of deactivation. Engstrom-Heg (1972) developed time-lapse curvilinear relationships for deactivating Noxfish® (5% rotenone) concentrations over a range of $KMnO_4$ concentrations in distilled water (Figure SOP 7.1). In distilled water, rotenone is deactivated with KMnO4 at approximately a 1:1 (ppm $KMnO_4$: ppm Noxfish) ratio, when the contact time ("cutoff" in the Figure SOP 7.1) is about 60 minutes. As the contact time is shortened, the ratio of $KMnO_4$: rotenone increases. For example, at 30 minutes contact time the ratio is 1.5–2.0:1.0. The 30-minute contact time is recommended for most stream treatments to assure deactivation. Contact periods shorter than 30 minutes are less efficient and problematic because greater amounts of $KMnO_4$ are required, resulting in higher levels of residual $KMnO_4$ that travel downstream from the Project Area. Contact periods greater than 30 minutes are problematic because it is difficult to maintain $KMnO_4$ residues in the water column without booster stations.

2. Another important consideration is that dissolved electrolytes, aquatic plants, dissolved and suspended organic matter, and streambed materials increase the amount of $KMnO_4$ required. These contribute to the background "oxygen demand" of water that typically ranges from 1–4 ppm for 30 minutes contact time. (e.g., water flowing over granitic bedrock surfaces may have an oxygen demand

of only 1 ppm, while in karst or limestone landscapes, streams with watercress and dissolved organics may have a demand of 4 ppm).

3. The ability to measure $KMnO_4$ in the stream is central to assuring that sufficient material has been applied and that rotenone is deactivated. More $KMnO_4$ than is necessary to deactivate rotenone is applied to yield a measurable $KMnO_4$ residual at the end of the 30-minute contact zone, and this way the Certified Applicator can be sure that all rotenone is gone at the end of the 30-minute contact time. For simplicity, it is recommended that a residual level of 1 ppm $KMnO_4$ be maintained at the end of the contact zone. This is a level not likely toxic to fish during short-term exposures, but can be measured by most commercial devices and is also easily visible to the unaided eye.

4. For example, the application rate of 4 ppm $KMnO_4$ is required for a 30-minute contact zone to deactivate 1 ppm Prenfish® in granitic soils (or equivalent 5% liquid formulation). The calculations are: 2 ppm $KMnO_4$ is needed to deactivate 1 ppm Prenfish® + 1 ppm $KMnO_4$ is needed for background oxygen demand in water + 1 ppm $KMnO_4$ residual is need at the 30-minute travel-time mark.

5. It is only through experience and knowledge of local conditions, soils and rock types that the Certified Applicator will be able to closely predict the oxidizing demand of a stream. In most circumstances, it will be necessary to use professional judgment in selecting an initial concentration.

B. Application Methods

1. Deactivation is a dynamic operation that presents some difficulties in managing application rates and predicting rotenone concentrations. There is no practical field procedure to determine actual rotenone (i.e., real-time) concentrations at the deactivation station. Certified Applicators should consider using a fully functional, redundant (i.e., backup) deactivation station positioned downstream from the effective reach of the primary deactivation station. The sensitivity of the public, the relative value of the aquatic community downstream of the project, and other issues will determine the need for redundant systems.

2. Some Certified Applicators begin deactivation procedures at the onset of rotenone application to assure that no rotenone passes the deactivation station and to reduce permanganate demand of the streambed immediately downstream from the deactivation station. Other Certified Applicators determine the time (sometimes with dye) when rotenone is expected to arrive at the deactivation station and begin application of $KMnO_4$ at that time. It takes a minimum of 1–2 hours of $KMnO_4$ application for most substances in the streambed to become oxidized and thus, there is less $KMnO_4$ available to reduce rotenone within the first several hours of operation. The deactivation system is operated prior to treatment for several hours during calibration exercises to assist in oxidizing streambed materials.

3. Measurements of residual $KMnO_4$ are taken periodically to assure the residual level of 1 ppm is present at the end of the 30-minute contact zone. Deviations can be addressed by increasing or decreasing the rate of application of $KMnO_4$ to the stream. Measurements every half hour are usually necessary at the beginning of

the treatment, but are scaled back to every 1–2 hours once equilibrium is achieved. Normally, the $KMnO_4$ demand within the deactivation zone decreases over the duration of the treatment as monitored by the residual $KMnO_4$ at the end of the 30-minute contact zone. It is beneficial to have the deactivation station in constant contact (e.g., two-way radios) with the person(s) doing the permanganate monitoring so adjustments can be made instantly, if necessary. The amount of $KMnO_4$ required will decrease as rotenone residues dissipate over time.

4. The effectiveness of the deactivation is measured by the ability of caged fish to survive in water downstream from the 30-minute contact zone. A 1 ppm residual $KMnO_4$ is not toxic to trout at exposures of less than 96 hours. Mortality of fish at this location is likely attributable to rotenone or other factors such as confinement stress. Replace sentinel fish daily where practical because of confinement stress in flowing water.

5. Continue deactivation until the water directly upstream of the deactivation station can sustain fish in an unstressed state in a bioassay for a minimum of 4 hours. Use caution when using the target fish species in a bioassay at this location unless the end of the deactivation zone is below a fish barrier, especially if this is the last of a series of treatments. Fish that escape cages will not compromise the success of the treatment if it is below a fish barrier.

6. Generally, caged fish are placed above the point of $KMnO_4$ injection and at the end of the 30-minute contact zone. Placing caged fish at the 15-minute contact time point may help with interpreting the progress of deactivation within the deactivation zone. Fish at 30 minutes will survive and those at 15 minutes will show signs of stress but will take many hours to die. Another location for caged fish is below the reach of the redundant (or backup) deactivation stations to provide assurance that total deactivation was achieved. This site is useful for interpretation purposes when fish below the 30-minute contact zone start to die.

7. Continual maintenance of the deactivation stations is required throughout the treatment. On streams, at a minimum expect to run this operation for the duration of the treatment plus the travel time from the most upstream point of project downstream to the deactivation station. For outlets from standing bodies of water, deactivation should continue until water is no longer toxic and may be prolonged by low water temperatures. A supervisor can provide assistance to stations as needed for breaks, errands, and loading and monitoring live fish cages. Each worker maintains a log of activities, application rates, and observations to ensure that prescribed procedures are followed and for future reference.

## C. Application Equipment

1. Dispense $KMnO_4$ in liquid or solid form. Liquid solutions should consist of 1 pound solid $KMnO_4$ mixed into 5 gallons (25 g/L) water. Most liquid applications use a reservoir with a metering device to dispense at a constant measurable rate. However, varying conditions during treatment (changing discharge rates, rotenone concentration, and oxygen demand) may require a change in the dispensing rate of $KMnO_4$, and use of a metering device that can be manually adjusted. A 2.5% $KMnO_4$ solution is dispensed at a constant concentration using the equation:

**LF = Y • 70 • Q** where, LF = flow of 2.5% $KMnO_4$ solution (ml/min), Y = desired $KMnO_4$ concentration in stream (ppm) and Q = stream discharge ($ft^3/s$) or **LF = Y • 1.982 • Q** where, LF = flow of 2.5% $KMnO_4$ solution (ml/min), Y = desired $KMnO_4$ concentration in stream (ppm) and Q = stream discharge ($m^3/s$). Application of solid $KMnO_4$ is typically done with a device that meters the material out of a reservoir (or hopper) directly into the water. Again, the metering device should be adjustable. $KMnO_4$ crystals can be added to the stream using a mechanical auger of other device using the equation:
**SF = Y • 1.7 • Q** where, SF = flow of solid $KMnO_4$ crystals (g/min), Y = desired $KMnO_4$ concentration in stream (ppm) and Q = stream discharge ($ft^3/s$) or **SF = Y • 0.048 • Q** where, SF = flow of solid $KMnO_4$ crystals (g/min), Y = desired $KMnO_4$ concentration in stream (ppm) and Q = stream discharge ($m^3/s$).

2. The advantage of liquid formulation application is that the devices require no electricity to run, and are very mobile; thus, these applications are well suited to remote locations with no power or road access. The disadvantage for liquid application is that it requires a large reservoir or making large amounts of liquid for streams with large discharges or if deactivation is extended for many days. The advantage for solid formulation application is that a large reservoir is not needed and the metering device is typically extremely accurate. The disadvantage for solid formulation application is that the equipment is heavy, unwieldy and needs electricity and thus best suited to areas with road access and deactivation of high discharge streams.

D. Measuring Potassium Permanganate

1. $KMnO_4$ can be measured directly in the field using several analytical techniques. The most accurate method is to make a standard spectrophotometric curve that plots absorbance (at 525 nm) for solutions of known $KMnO_4$ concentration. Samples taken in the field during treatment are measured for absorbance on a portable spectrophotometer, and concentrations can be estimated by reading off the standard curve. See Standard Method 4500-$KMnO_4$ B for details (American Public Health Association 1998).

2. A less accurate estimate of $KMnO_4$, but an easier approach, is the DPD (N, N-diethyl-p-phenylenediamine sulfate) method for measuring total chlorine. Through the introduction of a powder containing DPD, potassium iodine, and a buffer, the oxidizing potential of the solution is measured spectrophotometrically or using a color wheel. The oxidizing potential of permanganate is about 89% of total chlorine, so results from this method can be multiplied by 0.89 to get an approximate measure of $KMnO_4$ concentrations in water (see Hach Chemical Company Method 8167).

III. Additional Information

American Public Health Association. 1998. Standard Methods for the Examination of Water and Wastewater, 20th edition. American Public Health Association, Washington, D.C.

Engstrom-Heg, R. 1972. Kinetics of rotenone-potassium permanganate reactions as applied to the protection of trout streams. New York Fish and Game Journal 19(1):47–58.

Marking, L.L., and T.D. Bills. 1975. Toxicity of potassium permanganate to fish and its effectiveness for detoxifying antimycin. Transactions of American Fisheries Society 104:579–583.

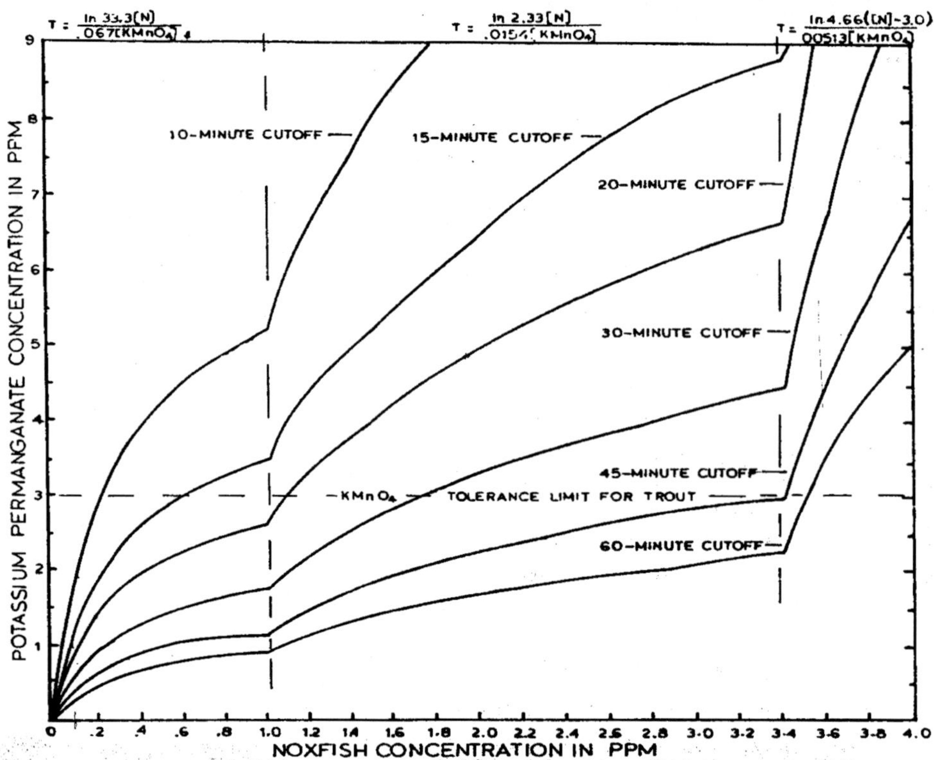


**Figure 5. Relationship between Noxfish concentration, potassium permanganate concentration and contact time required for detoxification.**

FIGURE SOP 7.1. Time required (contact-time) for potassium permanganate to deactivate rotenone from Engstrom-Heg (1972).

# Operation of Semi-Closed Probe Systems for Application of Liquid Rotenone Concentrate
# SOP:8.0

**PROCEDURE TITLE:** Operation of Semi-Closed Probe Systems for Application of Liquid Rotenone Concentrate

**APPLICABILITY:** Application of rotenone in waters of the United States

**PURPOSE:**

1. Mitigate for occupational exposure to liquid rotenone
2. Provide guidance on the design and operation of semi-closed probe systems for application of liquid rotenone concentrate typically used in large projects

**LOCATION ON LABEL:** "Hazards to Humans and Domestic Animals" under the heading "Precautionary Statements"

**PROCEDURE:**

The semi-closed probe system allows for the transfer of liquid rotenone concentrate directly from the drum to the mixing tank/application device with the use of a suction hose connected to one end of the suction pump on the mixing tank/application device and connected at the other end to a probe/dip tube. The rotenone concentrate is diluted and mixed in the pump head and discharged to site water without applicator contact.

I. Preparation

A. Stand drum with top end and bungs pointed up. Ensure that bungs are tight. Then tip drum over carefully and roll back and forth to suspend any settled material at the bottom of the drum.

B. Stand drum on the ground or a secure level platform with top end and bungs pointed up. Then remove 2″ or ¾″ plug from bung of drum.

C. Do not pour liquid rotenone directly from its drum.

II. Operation

A. Remove the 2″ or ¾″ plug from bung of drum and insert the probe/dip tube into the bung hole until the foam ring/gasket on the probe fits snugly around the bung hole to minimize leakage of liquid rotenone.

B. Size the metal or chemically-resistant plastic probe/dip tube for a snug fit into the bung when a foam ring/gasket is attached. The anti-drip flange of the barrel (curved inward towards bung) ensures that excess liquid rotenone returns to the barrel when the probe/dip tube is removed.

C. Attach a matching chemically-resistant rigid hose to the probe/drip tube to suck the liquid rotenone concentrate from the drum into the pump head for mixing with site water. Shut-off valves are located at the probe/dip tube end and at the pump end (see Figures SOP 8.1 and 8.2). These valves control the mix of rotenone liquid concentrate to site water (1 part rotenone concentrate: 9 parts site water (10%) is recommended dilution) for quick dispersion in standing waters. Calibrate the semi-closed delivery system before use.

D. Do not handle the probe/dip tube in a manner that allows dripping or splattering of the liquid rotenone onto yourself or any other person. Do not touch the portion of the probe/dip tube that has been in contact with this product until the probe has been triple rinsed with water.

E. Applicants using a boom or other mechanized equipment must discharge rotenone below the water surface.

F. Applicants using hand-held or hand-directed nozzles may release rotenone above the water surface. Operate hand-directed nozzle in a manner to minimize small droplets and reduce drift. A large volume of water relative to the amount of rotenone concentrate minimizes drift. A large diameter orifice on the spray nozzle, spraying with the prevailing wind, reducing the distance sprayed from the nozzle, and directing the spray towards the water surface all minimize drift.

G. Once the entire liquid rotenone has been removed from the drum, triple-rinse the probe while it remains inside the drum, if possible. This can be accomplished by adding site water through the other bung hole, either manually or using the water pump and another hose on the outlet structure (see Figure SOP 8.3). If an unrinsed probe must be removed from the drum, triple rinse it and all parts of the aspirator in treated site water.

H. Take the following steps if the probe/dip tube must be disconnected from the suction hose before both the probe and the hose have been triple-rinsed: (1) equip the probe end of the hose with a shutoff valve; (2) install a dry break coupling (see Figure SOP 8.4) between the valve and the probe, and; (3) close the shutoff valve before disconnecting the probe.

III. Safety

A. Mixers, loaders, and applicators using all systems must wear PPE as described on the product labeling.

B. All systems must be capable of removing the pesticide from the shipping container and transferring it into mixing tanks and/or application equipment.

C. At any disconnect point, the system must be equipped with a dry disconnect or dry couple shutoff device to minimize dripping (see Figure SOP 8.4).

## IV. Application

With properly operating semi-closed systems, the liquid concentrate is diluted in the pump head and boats can apply diluted rotenone underwater (see Figure SOP 8.5) or as a diluted spray using a hose and hand-directed nozzle (see Figure SOP 8.6).

## V. Equipment

### A. Mixing Equipment

1. Use a high-pressure (i.e., 60 to 70 psi) centrifugal self-priming water pump (2 to 3 inch) sufficient to pump water through a probe and create a vacuum to draw liquid rotenone from container. The pump functions as the 'mixing tank' to combine water from the intake hose and the liquid rotenone concentrate. Examples include Honda WH20XK1AC1 or Berkeley B1-1 pumps.

2. Use a high-pressure (i.e., 100 to 150 psi) water intake hose or plastic pipe. The intake hose/pipe is attached to a T-coupling having the liquid rotenone intake suction hose attached at the distal end (see Figure SOP 8.1). Screen the distal end of the water intake hose to reduce the amount of vegetation and/or debris entering the hose (see Figure SOP 8.2).

3. Attach a valve at the distal end of the liquid rotenone concentrate intake suction hose (i.e., ½ or 1½ inch diameter). The valve controls the rate of liquid rotenone flow into the water intake. The probe is inserted through a foam or rubber ring or gasket into the bung hole of the drum (see Figure SOP 8.2).

4. A water outlet hose (2 to 3 inches diameter) normally applies the diluted and mixed rotenone underwater (see Figure SOP 8.5) unless connected to a spraying device (see Figure SOP 8.6).

### B. Hand-directed nozzle

The dilute mixture of rotenone can be directed to a nozzle for spraying shallow or inaccessible waters (see Figure SOP 8.6). A brass straight stream nozzle is attached to a rotating monitor, which is mounted to the boat deck or cowling (see Figure SOP 8.7). A shut-off valve is between the outlet hose/piping and the nozzle.

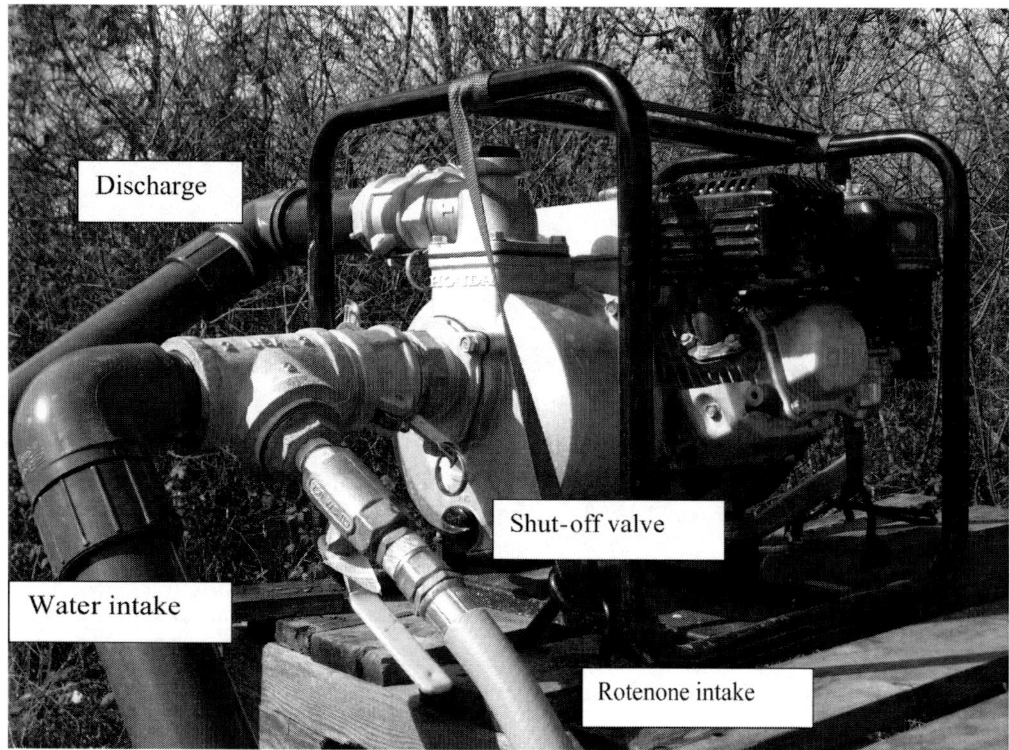


FIGURE SOP 8.1. Two-inch gasoline-powered water pump showing arrangement of intake and discharge hoses and shut-off valve (Photo credit, Holly Truemper, Oregon Department of Fish and Wildlife).

FIGURE SOP 8.2. Shut-off valves on both ends of rotenone liquid concentrate probe/dip tube (Photo credit, Holly Truemper, Oregon Department of Fish and Wildlife). This picture does not show the now required foam gasket on the probe/drip tube.

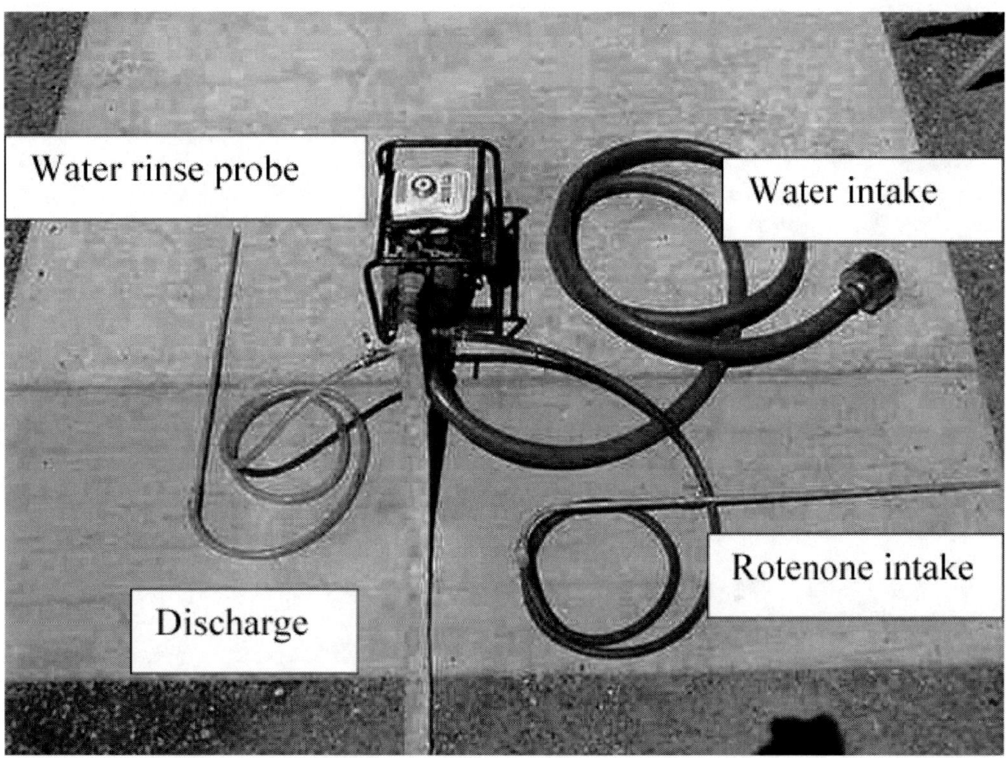


FIGURE SOP 8.3. Arrangement of site water intake and discharge hoses and rotenone liquid concentrate probe/drip tube (foam gasket not shown) and optional site water rinse probe (Photo credit, Dave Rose, California Department of Fish and Game).

FIGURE SOP 8.4. An example of a dry disconnect valve showing the two halves coupled together with cam arms. Two interlocking handles make opening and closing easy. No guessing involved. When liquid is flowing, disconnect valve halves cannot be uncoupled without turning handles to closed position. When disconnect valve halves are apart, handles cannot be turned to open position.

FIGURE SOP 8.5. Application of liquid rotenone below waterline using semi-closed application system at Diamond Lake, Oregon (Photo credit Brian Finlayson, California Department of Fish and Game).

FIGURE SOP 8.6. Application of liquid rotenone using semi-closed application system with spray nozzle to shallow waters of Park Lake, Grant County, Washington (Photo credit, Jeff Korth, Washington Department of Fish and Wildlife).

FIGURE SOP 8.7. Examples of monitors (fire fighting equipment) and suggested nozzle.

# Operation of Semi-Closed Aspirator Systems for Application of Powdered Rotenone
# SOP:9.0

**PROCEDURE TITLE:** Operation of Semi-Closed Aspirator Systems for Application of Powdered Rotenone

**APPLICABILITY:** Application of rotenone in waters of the United States

**PURPOSE:**
1. Mitigate for occupational exposure to powdered rotenone
2. Provide guidance on the design and operation of semi-closed probe systems for application of powdered rotenone for fish control

**LOCATION ON LABEL:** "Engineering Controls for Mixing/Loading for Wettable Powder Formulations" under the subheading "Hazards to Humans and Domestic Animals" under the heading "Precautionary Statements"

**PROCEDURE:**

The semi-closed system sucks powdered rotenone directly from the product container, mixes it with the water from the treatment site in the aspirator, and then discharges the slurry at or slightly below the waterline. A high-pressure pump forces water through the aspirator creating sufficient suction to vacuum powdered rotenone from containers. Discharge slurry directly to the water at or below surface or above the surface through a hand-directed nozzle as a spray. The semi-closed aspirator system significantly reduces airborne rotenone dust.

I. Preparation

A. Place powdered rotenone container on an impermeable surface, such as a tarp or concrete ramp at the water's edge, so spillage can be safely flushed into the treated water.

B. Roll the sealed container on the ground to loosen the rotenone powder that may have settled during shipping and storage.

C. Remove the 2" or ¾" plug from bung of container liner enclosing this product when container is sitting on the ground or on a secure level platform, with the bung end of the container pointed up.

D. Do not pour powdered rotenone product from its container.

II. Operation

A. Start the pump and establish water flow through the aspirator to achieve Venturi suction.

B. Size the metal or chemically-resistant plastic probe/dip tube for a snug fit into the bung when a foam ring/gasket is attached. The plastic liner has an anti-drip flange (curved inward towards bung) to remove excess powder when the probe/dip tube is removed.

C. Remove top of the container, remove plug from bung of the plastic liner and insert probe until the foam ring/gasket on the probe fits snugly around the bung opening in the plastic liner.

D. Transfer product from the container with the use of the probe/dip tube attached to the suction hose through the top of the aspirator. The aspirator is connected to the discharge side of the suction pump.

E. Maneuver the probe/dip tube by hand to suction powdered rotenone from container. Handle the probe/dip tube in a manner that minimizes the dispersing of rotenone powder onto you, any other person, or into the air. Know the rate of dispersal to ensure that the required amount of rotenone powder is dispensed evenly across the treatment area.

F. Remove the probe/dip tube when the entire contents of the container have been removed. Let the probe/dip tube suck air for 30 seconds prior to applying powdered rotenone from the next container.

G. Do not attempt to triple rinse the probe/dip tube and suction hose with water until the entire application has been completed. Make sure equipment is completely dry after cleaning before reusing. A wet/damp probe/dip tube and suction hose will lead to the powder caking and reducing the suction capacity of the aspirator system.

H. After the application is complete and the aspirator probe is removed from container, shake residual powder into bottom of plastic liner, fold the plastic liner into the container, and reseal the container.

I. Liners are triple-rinsed by removing from the container, submerging the liner before cutting open, and wetting the liner underwater. Conversely, a wash-down pump and line may be used to safely and effectively triple-rinse the liners and containers. The wash line should be inserted into the bung hole, wetting the powder, and the slurry can then be rinsed directly into the treated water.

J. Following treatment, deployment equipment should be emptied, triple-rinsed with water to remove residue, disassembled, collected and stored in accordance with the product label. Clean equipment where rinse water can be effectively disposed into the treatment site. Make sure all equipment especially the aspirator and suction hose is completely dry prior to the next application. The operation of the aspirator can be checked with dry, fine grain sand.

III. Safety

A. Personal Protective Equipment: mixers, loaders, applicators and other handlers must wear PPE as required on product labeling (see Figure SOP 9.1).

B. All systems must be capable of removing the powder from the shipping container for application and transferring it into mixing tanks and/or application equipment.

IV. Application

With properly operating semi-closed systems, boats can apply powdered rotenone efficiently (see Figure SOP 9.2).

V. Equipment (Figure SOP 9.3)

A. Examples of pumps that have worked well for project personnel: Gorman-Rupp series 60 centrifugal model with an enclosed impeller, or equivalent, coupled to a Briggs and Stratton twin cylinder, 18 HP, air cooled gasoline engine. Pump is rated at 150 GPM at 65 feet of head.

Hale transportable fire pump coupled to a Briggs and Stratton Vanguard series 350400 V Twin, overhead valve, air cooled design, rated at 18 BHP at 4000 rpm with a torque of 30 lb-ft at 2600 rpm. Pump is rated at 150 GPM at 100 psi.

B. Hoses—Water intake and slurry delivery hose is high-pressure hose, rated at 100 to 150 psi, recommended for delivering water from the pump to an aspirator. Use a screen at the distal end of water intake hose to reduce the amount of vegetation and/or debris entering the hose. Use a weight attached to the distal end of the water intake hose to keep hose in the water.

C. Aspirator—Aspirator is manufactured with readily available plumbing connectors and pipe (Figure SOP 9.4). Mount aspirator directly to pump outflow for best suction through powder suction probe/dip tube and efficient mixing of rotenone and water slurry.

VI. Additional Information

Finlayson, B. J., R. A. Schnick, R. L. Cailteux, L. DeMong, W. D. Horton, W. McClay, C. W. Thompson, and G. J. Tichacek. 2000. Utah's rotenone aspirator system. Appendix L in Rotenone use in fisheries management: administrative and technical guidelines manual. American Fisheries Society, Bethesda, Maryland.

Thompson, C. W., C. L. Clyde, D. K. Sakaguchi and L. D. Lentsch. 2001. Utah's procedure for mixing powdered rotenone into a slurry. Pages 95–105 *in* R. L. Cailteux, L. DeMong, B. J. Finlayson, W. Horton, W. McClay, R. A. Schnick, and C. Thompson, editors. Rotenone in fisheries: are the rewards worth the risks? American Fisheries Society. Trends in Fisheries Science and Management 1, Bethesda, Maryland.

FIGURE SOP 9.1. Powered air-purifying respirators used in applying powdered rotenone to Quincy Lake, Washington (Photo credit, Jon Anderson, Washington Department of Fish and Wildlife).

FIGURE SOP 9.2. Application of powdered rotenone using semi-closed aspirator system to Diamond Lake Oregon (Photo Credit, Brian Finlayson, California Department of Fish and Game).

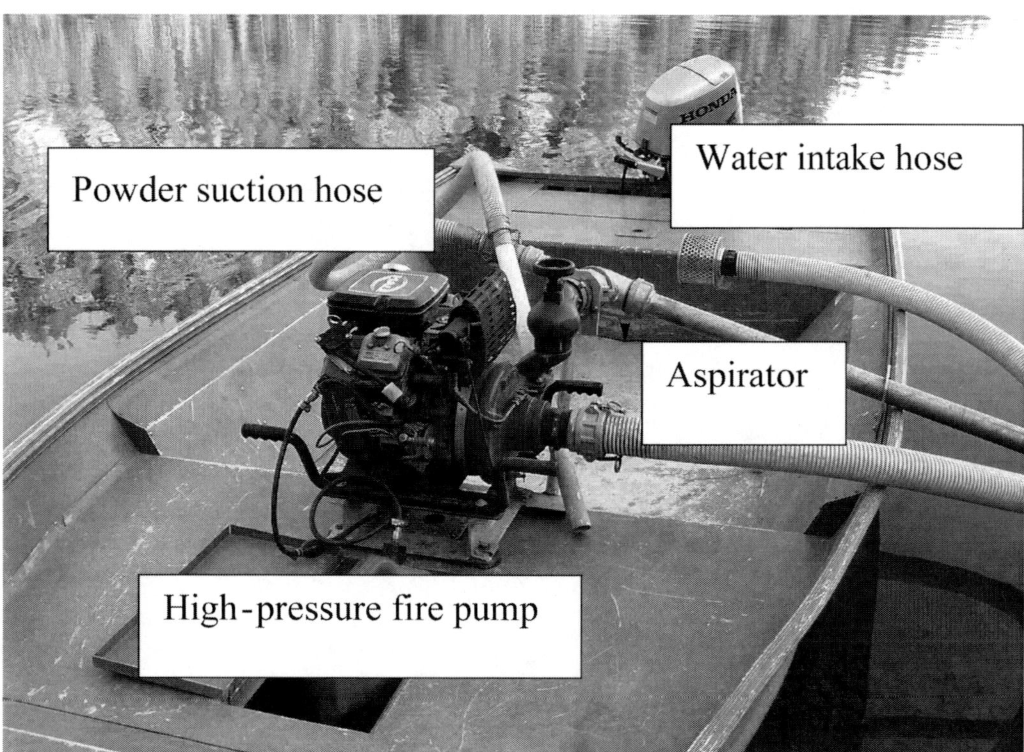


FIGURE SOP 9.3. Gasoline-powered water pump showing arrangement of intake and discharge hoses (Photo credit, Jon Anderson, Washington Department of Fish and Wildlife). Note: the powder suction hose is not equipped with the now- required foam gasket.

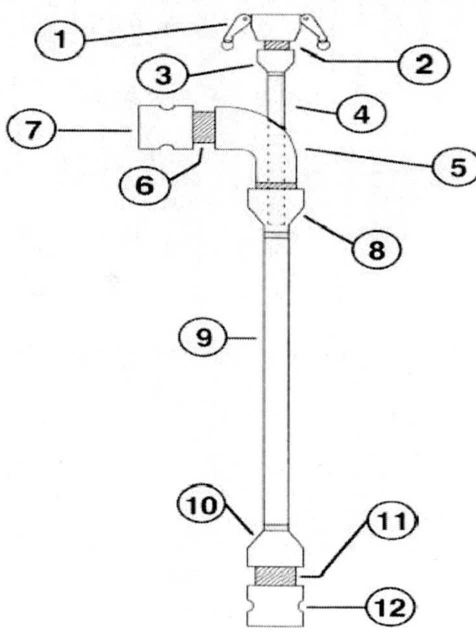


Figure SOP 9.4. Rotenone aspirator developed by Utah Division of Wildlife Resources: (1) 2" female camlok fitting (vacuumed powder enters), (2) 2" x 2" nipple, (3) 2" x 1.25" bell reducer, (4) 1.25" x 12" nipple, (5) 3" 90 degree street elbow, (6) 3" x 3" nipple, (7) 3" male camlok fitting (lake water enters), (8) 3" x 1.5" bell reducer (powder and water mix) , (9) 2" x 48" galvanized pipe, (10) 2" x 3" bell reducer, (11) 3" x 3" nipple, (12) 3" male camlok fitting (powder-water slurry discharged). Rotenone aspirator was constructed of galvanized pipe (from Thompson et al. 2001).

# Transferring (Mixing/Loading) Liquid Rotenone Concentrate
# SOP:10.0

**PROCEDURE TITLE:** Transferring (Mixing/Loading) Liquid Rotenone Concentrate

**APPLICABILITY:** Application of rotenone in waters of the United States

**PURPOSE:**

1. Mitigate for occupational exposure to liquid rotenone
2. Provide guidance on transferring (mixing and loading) liquid rotenone concentrate from product containers to service containers and application equipment

**LOCATION ON LABEL:** "Transferring (Mixing/Loading) Liquid Formulations" under the subheading "Hazards to Humans and Domestic Animals" under the heading "Precautionary Statements"

**PROCEDURE:**

I. Precautions

A. Preparations

Do not open pesticide containers until ready for use. All personnel involved with the application will be knowledgeable about rotenone's toxicity, the product label, MSDS, spill contingency plan and any site specific safety plans. All personnel coming in contact with rotenone must wear proper personal protective equipment including proper clothing, eye protection and dust/mist respirators. Rotenone applications require pouring, diluting, pumping, mixing and transferring the liquid formulation. Good pesticide handling practices minimize even minor spills, which lead to clean, safe and efficient rotenone treatments (see SOP 3). Many, if not most, minor liquid rotenone spills occur when transferring rotenone formulations from one container to another. Other minor spills occur when open containers are tipped, etc. Mixers, loaders, and applicators are required to wear PPE as specified on product label.

B. Secondary Containment

Control minor spills by transferring rotenone concentrate from original containers to service containers (see SOP 4 for requirements) or application equipment within a secondary containment area. Plastic-lined berms, tubs or stock tanks are effective means of secondary containment. Consistent use of secondary containment results in fast and complete recovery of small quantities of spilled materials that are used at the treatment site.

To create a bermed area, soil or other suitable material (e.g., hay bales) can be shoveled, mounded or otherwise fashioned in a complete circle and then covered with a one-piece plastic liner capable of containing any and all rotenone. Consult the responsible land-use agency if soil excavation is required. If a spill occurs, immediately recover the spilled material by sponge, pump or other efficient means consistent with spill contingency plan (see SOP 4). If appropriate, apply the spilled material to the treatment area.

Wash the secondary container or plastic liner, original container, instruments used in transfer, and measuring device with site water and dispose of the rinseate into treated site water. Carefully fold the plastic liner upon itself, being careful not to release any residual material and wash liner in the water of the treatment area.

C. Transfer

Staff involved with mixing and loading should transfer product from original containers to service containers or application equipment by measuring appropriate amounts into measuring devices (i.e., measuring cups or graduates) and then transferring the measured rotenone into a service container or application equipment. When application equipment is not loaded with rotenone concentrate within the secondary confinement area, transfer the rotenone concentrate from the service container to application equipment at the application site. Do not handle this product in a manner that drips or splatters the product onto yourself or any other person.

II. Rotenone Concentrate from Product Containers ≤ 5 Gallons

Transfer rotenone concentrate from 1- and 5-gallon product containers within secondary confinement area. Pouring the product from the original container into a measuring devise is allowed. Small pumps or pipette-type devices can also transfer rotenone from the original container to measuring devices. Measuring devices are a non-porous receptacle marked with the appropriate graduations (i.e., gallons, liters, ounces, milliliters, etc). Containers remain within the secondary containment during transfer.

Alternatively, 5-gallon containers may have threaded bungs that can accept a faucet. Remove the factory bung and outfit the bung hole with a conventional faucet and place the container on its side on a rack fabricated for this purpose. Place the rack and drum inside of a secondary containment area and fill measuring devices by means of the faucet. It may be necessary to drill a small hole in the top of the drum to allow air to enter the container. This may affect the ability to recycle the 5-gallon drum and may cause storage problems with any unused portion of rotenone concentrate.

III. Rotenone Concentrate from Product Containers > 5 Gallons

Product label does not permit one to pour rotenone concentrate from containers > 5 gallons. Transfer liquid rotenone concentrate from original container to a measuring device within a secondary containment area described above using small hand or electric drum pumps for this purpose. This will allow applicators to better control liquid transfer and eliminate minor spills. Use of such pumps can also cut down on physical strain from lifting and holding heavy containers. Spilled material can be easily recovered from the secondary containment area.

## IV. Spill Containment

In the event that a spill occurs, it is of paramount importance that the spilled material be contained. See SOP 4 for information on Spill Prevention and Containment and a Spill Contingency Plan.

# Operation of Drip Stations, Peristaltic Pumps and Propwash Venturi for Application of Liquid Rotenone SOP:11.0

**PROCEDURE TITLE:** Operation of Drip Stations, Peristaltic Pumps and Propwash Venturi for Application of Liquid Rotenone

**APPLICABILITY:** Application of rotenone in waters of the United States

**PURPOSE:** Provide guidance on application of liquid rotenone to streams and rivers (drip stations and peristaltic pumps) and lakes and ponds (propwash venturi).

**LOCATION ON LABEL:** "Directions for Use"

**PROCEDURE:**

I. Preparation and Safety

A. Applicators of liquid rotenone must wear PPE including coveralls or long-sleeved shirt and long pants, chemical-resistant gloves, chemical resistant footwear plus socks, protective eyewear, and a dust/mist respirator. Waterproof waders may be worn in place of the chemical-resistant footwear. PPE is required when manipulating or adjusting the equipment or when in contact with treated water.

B. Liquid rotenone is typically transferred from the product container to a measuring device and then to a service container or application device inside of a plastic-lined bermed or otherwise self-contained area (see SOP 10). Service container labels must identify the following information (1) the name and address of the person or firm responsible for the container, (2) the identity of the pesticide in the container, and (3) the signal word "Danger," "Warning," or "Caution," in accordance with the label on the original container. (see SOP 4). Typically, the transfer from manufacturer's container to service containers or application equipment occurs at some centralized location. However, drip cans, pumps and service containers can be taken to the streamside application site for transfer.

II. Drip Cans

A. Construction

Drip cans consist of a reservoir and a delivery apparatus for the application of rotenone to flowing waters. The reservoir can be any size, made from any material but typically plastic or metal in construction, fashioned from commercially available buckets or cans, and hold from 3 to 10 gallons liquid. The delivery system provides regulated flow of rotenone to the stream for maintaining a constant concentration of rotenone. The best delivery system is metered, allowing for manual adjustment of the drip rate. Three designs are commonly used (see SOP 11 Appendix A).

B. Operation

Rotenone is applied continuously in small amounts for extended periods of time when treating flowing waters. Non-mechanized drip stations are commonly used to accomplish this. Many agencies and organizations have developed drip stations that are designed to provide constant feed rates so that treatment concentrations will be uniform. All drip cans minimize variations in flow rate attributable to loss of head pressure due to the decreasing amount of rotenone in the container. All drip cans are subject to some variation in output and periodic rate checks and adjustment (i.e., usually 30-minute intervals) are essential to successful use. The flow rate is checked by use of graduated cylinder or other measuring device and a stop watch.

C. Placement of Application Sites

Apply rotenone as a drip for 4 to 8 hours to flowing water. Multiple application sites are necessary along the length of the treated water to maintain the desired rotenone concentration (see SOP 5 for treatment rates). Application sites are generally spaced at no more than 2 hours or at no less than 1-hour travel-time intervals or based on bioassay results. Typically, these conditions result in spacing the application sites approximately 1/2 to 2 miles apart depending on water flow travel-time, stream gradient, solar radiation, turbidity, and other factors affecting rotenone degradation. A non-toxic dye like Rhodamine WT or Fluorescein is used to determine travel time.

D. Quantity of Rotenone

The amount of rotenone needed depends on many things including the desired in stream concentration of rotenone, the treatment duration, and the discharge of the receiving water (see Table SOP 11.1) using the equations:

**$X = F(1.692 \bullet C)$** where, X − ml/minute of undiluted rotenone formulation, F = flow of stream in $ft^3/s$, and C = desired rotenone formulation in ppm in stream or

**$X = F(0.04791 \bullet C)$** where, X = ml/minute of undiluted rotenone formulation, F = flow of stream in $m^3/s$, and C = desired rotenone formulation in ppm in stream.

For flows over 25 $ft^3/s$ (0.71 $m^3/s$), it is usually desirable to treat using undiluted rotenone formulation. For flows less than 25 $ft^3/s$, it is usually desirable to treat using diluted formulation.

Larger reservoirs (or re-filling of small reservoirs) are needed for treatments of long duration, greater rotenone concentrations, or larger streams and rivers. In practice, many applicators have reservoirs that are reused from one treatment to another. Because of this it is often desirable to either refill reservoirs as the situation dictates, or apply undiluted product.

TABLE SOP 11.1. The amount of undiluted liquid rotenone (ml) needed to achieve 1 ppm formulation concentration for 4-hour, 6-hour, and 8-hour treatments.

| Stream Discharge ($ft^3/s$) | Stream Discharge ($m^3/s$) | 4-h Treatment Rotenone (ml) | 6-h Treatment Rotenone (ml) | 8-h Treatment Rotenone (ml) |
|---|---|---|---|---|
| 1 | 0.0283 | 409 | 613 | 818 |
| 2 | 0.0586 | 818 | 1226 | 1635 |
| 3 | 0.0849 | 1223 | 1834 | 2445 |
| 4 | 0.1133 | 1631 | 2447 | 3263 |
| 5 | 0.1416 | 2040 | 3060 | 4080 |
| 6 | 0.1699 | 2449 | 3673 | 4898 |
| 7 | 0.1982 | 2858 | 4287 | 5715 |
| 8 | 0.2265 | 3263 | 4894 | 6525 |
| 9 | 0.2549 | 3671 | 5507 | 7343 |
| 10 | 0.2832 | 4080 | 6120 | 8160 |

The rotenone is dispensed from a container at a constant rate determined by the equation:

**Y = V ÷ T** where, Y = discharge rate of container (ml/min), V = container volume (ml) and T = treatment period of 240 min (4-h) to 480 min (8-h).

## III. Peristaltic pumps

Peristaltic pumps (9 or 12-volt) may be used in place of non-mechanized drip stations and offer advantages in certain applications. The advantages of the pumps include very stable pump rates and the ability to accommodate a wide range of discharge rates. These pumps will have the most utility where transport into remote areas is not problematic.

### A. Construction

The pumps are light and portable, but the batteries used for power are heavy. Masterflex (obtained through Cole-Parmer) and Control Company (obtained through United States Plastic Corp.) brand pumps, have been used with good success; however other brands and sources are available.

Flow rates are controlled by adjusting the pump speed setting, by the selection of one of several available hose diameters, and by selection of a high- or low-capacity pump (Masterflex) or different diameter nipples on the end of the tubing (Control Company). By varying these three components, steady flow rates from 5 ml/min to 1400 ml/min. can be achieved with the Masterflex pump or 0.4 to 85 ml/min with the medium-flow Control Company pump. One fully charged 12-V battery will power the Masterflex Model 7518-10 pump for over 12 hours at medium feed rates. Similar pump longevity can be achieved with a 9-V battery and the Control Company pump.

B. Operation

For the Masterflex pump, the rotenone container is placed inside a secondary containment vessel alongside the stream to be treated (see Figure SOP 11.1). The pump is set up above and adjacent to the rotenone supply. Pump capacity, hose size and pump setting are all predetermined matching the desired feed rate and pre-constructed output tables. A weighted pump intake hose is placed inside the opened rotenone container and the discharge is suspended over the stream. Pre-made tripods can be used for suspending the output hose. Power is provided by a simple wiring harness connecting the pump to the battery. For the Control Company pump, the battery and pump are placed on top of the rotenone container (See Figure SOP 11.2).

The flow rate is checked by use of graduated cylinder or other measuring device and a stop watch. The flow rate is corrected by adjusting the pump speed. If flow adjustment greater than what can be achieved by changing the pump speed is needed, changing the hose diameter may be required. For the smallest of streams where flow rates less than 5 ml/min of rotenone formulation are required, the rotenone should be diluted prior to application into the stream. Once a steady flow rate has been achieved, flow checks and adjustments should be made every 30–60 minutes.

C. Placement of Application Sites

Use the drip station placement strategy for pump application sites.

D. Quantity of Rotenone

As with drip cans, the amount of rotenone needed depends on many things including the desired instream concentration of rotenone, the treatment duration, and the discharge (see Table SOP 11.2) of the receiving water.

TABLE SOP 11.2. The dispensing rate of undiluted liquid rotenone (ml/min) needed to achieve a 1 ppm formulation concentration.

| Stream Discharge ($ft^3/s$) | Stream Discharge ($m^3/s$) | Flow of Undiluted Rotenone (ml/min) |
|---|---|---|
| 10 | 0.283 | 16.92 |
| 20 | 0.586 | 33.84 |
| 30 | 0.849 | 50.76 |
| 40 | 1.133 | 67.68 |
| 50 | 1.416 | 84.60 |
| 60 | 1.699 | 101.52 |
| 70 | 1.982 | 118.40 |
| 80 | 2.265 | 135.36 |
| 90 | 2.549 | 152.28 |
| 100 | 2.832 | 169.20 |

## IV. Venturi Boat-Bailer Systems

### A. Construction

This system consists of a reservoir and a delivery system for the application of dilute liquid rotenone to standing waters. The reservoir is typically a hard plastic tank that contains a pre-mixed dilution of the liquid rotenone formulation (1:10 rotenone:water is recommended). The reservoir has an air vent system to allow the product to flow freely. The orifice of the reservoir must be fitted with a shut-off valve which is attached to one end of a hose which extends to the venturi device. The venturi device is essentially an adjustable bracket which is clamped to the cavitation plate of the lower unit of an outboard motor. This bracket also has female threading to receive the male end of the hose extending from the reservoir. There is currently no known commercial source for these devices, so it will be necessary to work with a local machinist to provide a custom fabrication.

### B. Operation

The simplicity of the venturi system is that no pumps or electricity are required for its operation. The formulation flows under the force of gravity from the reservoir to the venturi bracket and out into the propwash (see Figure SOP 11.3). It is important to have these mounted so that the rotenone product is dispensed under the water surface and is directed into the propwash for mixing with the receiving waters. Typically, the venturi can be mounted to the cavitation plate on the lower unit of the outboard motor (see Figure SOP 11.4).

The flow rate from the reservoir tank is controlled by the shut-off valve. The applicator can adjust the flow rate so that the reservoir will be dispensed over a pre-determined period of time by considering the speed of travel and the total distance traveled while dispensing the contents of the reservoir. On a small lake, the path of the boat may be a back and forth pattern from one shoreline to another working along the long-axis of the lake. For a round lake, the path may be increasingly smaller concentric circles beginning at the lake margin and working toward the center. On large lakes, it is typical to divide the lake surface into treatment zones, with boats assigned to individual zones. These zones can be identified as actual topographic coordinates and a boat can stay within their zone with the use of a GPS receiver.

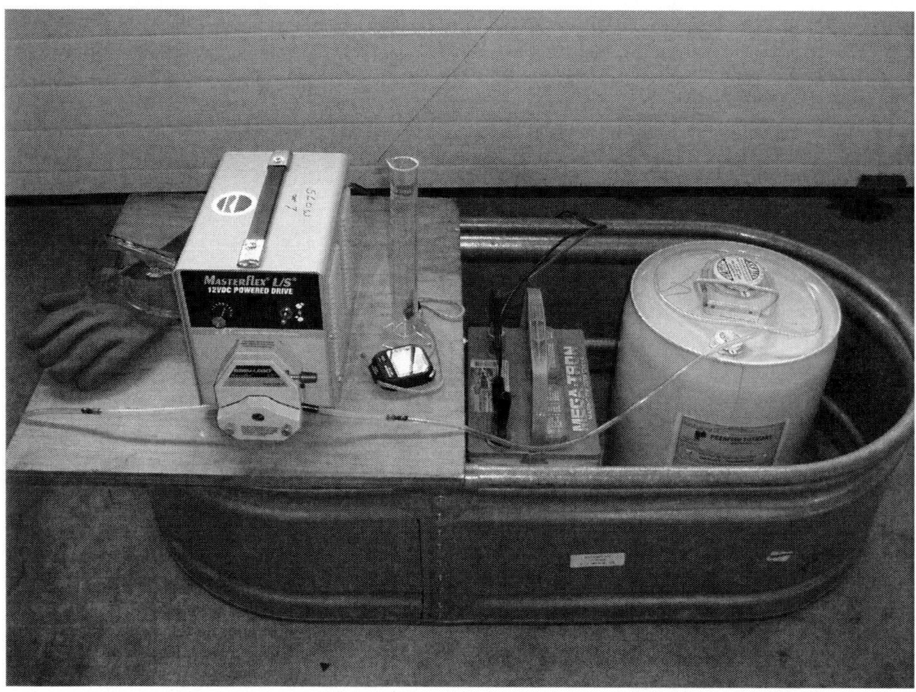


FIGURE SOP 11.1. Masterflex peristaltic pump set up in secondary containment area. Personal protective equipment, stop watch, and graduated cylinder are also shown.

FIGURE SOP 11.2. Control Company peristaltic pump placed on rotenone bucket showing dispensing tube into stream. A 9-V battery sits alongside the pump.

FIGURE SOP 11.3. Set-up of reservoir with pre-diluted liquid rotenone and venturi on small boat with out board motor.

FIGURE SOP 11.4. Close-up view of venturi attached to cavitation plate of an out board motor and tightened down using wing screws for quick installation/removal.

# Appendix A
# Design and Construction of Drip Cans

## Design A

1. Principle—A hose attached at one end to an orifice of a reservoir (which is sealed but vented per the Marriott bottle principle) and at the other end has a spigot for controlling drip rate. Once the spigot is adjusted to provide the desired drip rate, the constant vacuum in the sealed reservoir provides for constant drip regardless of quantity of material in the reservoir. This system allows for endless dispensing rates. The construction is described below.

2. Materials:

   ¼ inch copper tubing (approximately five feet)
   ¼ inch (inside diameter) x 3/8 inch (outside diameter) vinyl tubing (5-10 feet)
   ¼ inch compression brass needle valve (Lincoln Products #127410, Industry, CA 91740)
   5-gallon tin square can with screw cap (Freund Containers #1953, Chicago, IL 60620)
   (2) 7/32 inch x 5/8 inch stainless steel clamps
   Size 11 black rubber stopper with two predrilled holes

3. Procedure:

   A. Put stopper in can opening for measurements.

   B. Cut 2 pieces of ¼ inch copper tubing to length. Piece A should be ½ inch above bottom of can to two inches above the stopper (about 16 inches). Piece B should be 1 inch above bottom of the can to one inch above the stopper (about 14½ inches).

   C. Insert copper tubing into stopper holes so that Piece A is 2 inches above top of stopper and Piece B is 1 inch above top of stopper. Make sure that bottom of Piece B is about ½ inch above bottom of Piece A (see Appendix Figure 1).

   D. Use one of the stainless steel clamps to clamp the vinyl tubing onto Piece A above the stopper.

   E. Cut two, 4 inch long pieces of ¼ inch copper tubing. Attach the pieces into either end of the brass needle valve using the compression fittings.

   F. To one end of the needle valve, attach the other end of the vinyl tubing using the other stainless steel clamp. Note: The length of the vinyl tubing between the can and the valve is solely dependent on the reach between the can and the application site required for the treatment. Bend the other end of the copper tubing to function as a spout.

   G. To get the drip can working (see Appendix Figure 2):

      1. Fill the drip can with fluid.
      2. Insert the stopper into top of can.
      3. Open needle valve all of the way.

4. With finger over Piece B above the stopper, turn can on its side until fluid runs freely through the valve to the spout.
5. Turn can upright, take finger off of Piece B, and calibrate flow using needle value.
6. Make sure that the can is placed high enough above application site to provide adequate head for the entire treatment, the more head the better.
7. It may take several minutes for the flow to stabilize after adjustment.

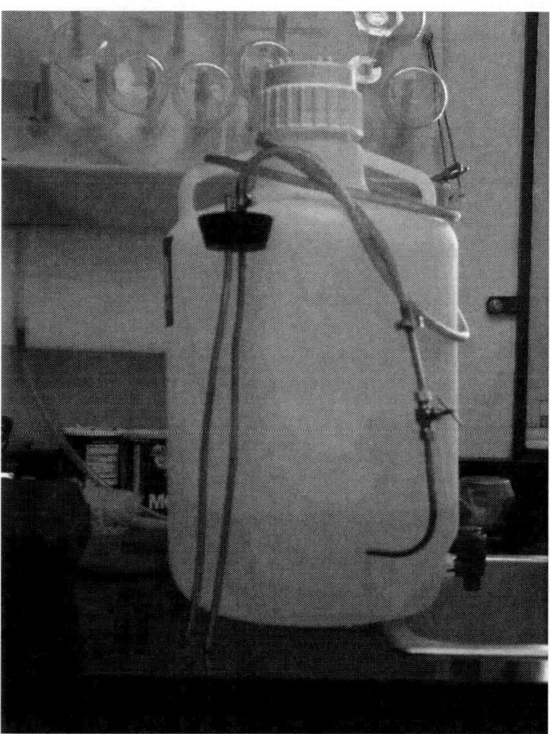

Appendix Figure 1. Copper tubes in rubber stopper.

Appendix Figure 2. An operating drip can.

**Design B**

1. Principle—A reservoir (3.5-gallon bucket with lid) feeds rotenone through a hose (see Appendix Figure 3) to a smaller reservoir which uses a float system (as in a toilet) to control head (see Appendix Figure 4). A hole is drilled in the bottom of this smaller reservoir to provide the desired drip rate. Drip rate can be changed with this apparatus by drilling a new hole or adjusting the operation of the float that changes the depth of fluid in the bowl enough to affect the head pressure and slow down or speed up the drip rate. This float system is described below.

2. Materials:

   3.5-gallon bucket with lid (Lab Safety Supply)
   Hull fitting, ¾ inch L (Cabela's Item # IE-012946)
   Garden hose male thread
   4 feet of ½ inch hose, cut-off valve, cone filter
   6 inch funnel
   1 Red Devil, 1-gallon nylon bag strainer
   Farnam Automatic Dog Waterer (Pet Vet Supply 1-800-283-2353)
   Hose clamp
   #9 cap thread gasket
   PVC couplings (1½ inch clean-out with threaded plug and 1½ inch adapter)

3. Procedure:

   A. With circle bit (i.e., key hole saw), cut 1½ inch hole in center of lid and affix PVC couplings (no cement needed).

   B. With circle bit, cut 1 inch hole in bottom of bucket.

   C. Attach hull fitting with #9 cap thread gasket on inside of bucket.

   D. Attach about 4 inches of ½ inch garden hose to the hull fitting and tighten with hose clamp.

   E. Attach a male thread fitting to the other end of the 4 inch piece of hose.

   F. Attach cut-off valve to male thread.

   G. Insert cone filter into female receptor on dog watering bowl, then attach female end of garden hose into cut-off valve and male end into dog watering bowl.

   H. Drill hole in center of dog watering bowl to provide for correct drip rate. A hole diameter of 0.059 inch (#53 drill bit) drips at a rate of 63 ml/min.

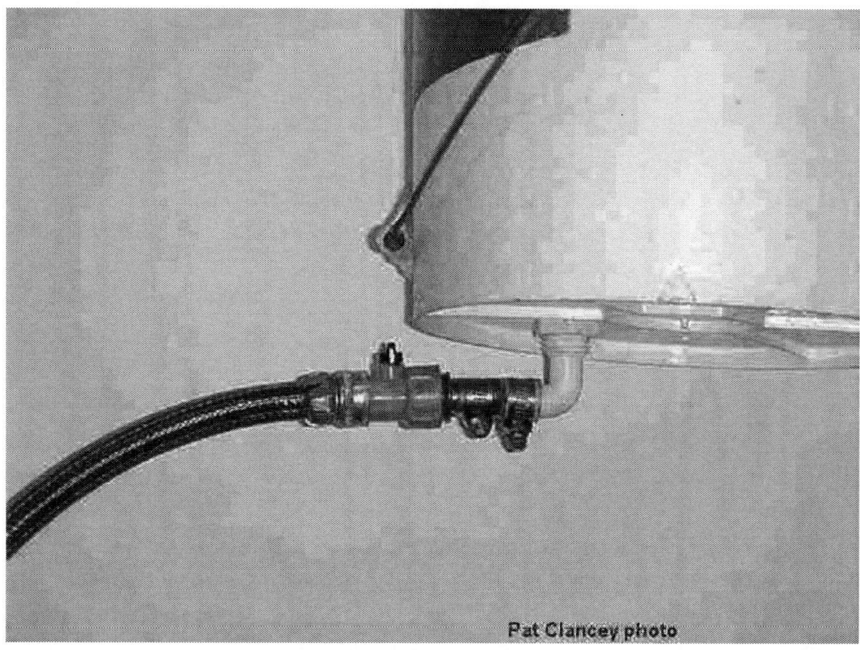


Appendix Figure 3. Drip bucket reservoir showing cut-off value and garden hose attachment.

Appendix Figure 4. Dog watering bowl reservoir and float in operation.

**Design C**

1. Principle—a reservoir utilizing the Marriott bottle principle but the pressure-equalizing vent is attached directly to the orifice and there is no spigot (see Appendix Figures 5 and 6). With this system, the drip rate is determined by the size of a hole drilled into the base of the T-shaped vent. Drip rate with this system can only be changed by drilling a new hole. This T-vent system is described below.

2. Materials:

   Coleman® 5-gallon water jug
   2-6 inch metal hose clamps
   Size 0 rubber stopper
   Pipeconx® (Uniseal®) flex coupling (1½ inch diameter x 3½ inch in length).
   PVC reducing bushing (½ inch x 1½ inch)
   PVC tee (¾ inch x ¾ inch x ½ inch (threaded))
   PVC slip cap (¾ inch)
   PVC pipe (¾ inch), cut two pieces, 8 inch and 2 inch in length
   PVC nipple (½ inch x 1½ inch length, outside thread on both ends)
   PVC primer and cement
   Teflon tape
   2- scrub pads (4 inch x 5¾ inch)

3. Procedure:

   A. Remove threaded water spout and air hole cover on new Coleman 5-gallon jug.

   B. Place Pipeconx flex coupling over water spout opening and rubber stopper in air hole.

   C. Glue 8 inch and 2 inch pieces of PVC pipe into either end of PVC tee.

   D. Glue cap over 2 inch piece of PVC and drill hole in cap. A #53 drill bit should yield a hole which flows at 75 ml/min.

   E. Screw ½ inch threaded nipple into short end of tee, using Teflon tape.

   F. Screw other end of ½ inch nipple into 1½ inch x ½ inch reducing bushing, using Teflon tape for good seal.

   G. Insert 1½ inch x ½ inch reducing bushing into flex coupling. Tighten hose clamps.

   H. Glue scour pads to underside of 5-gallon jug to prevent slippage while in operation.

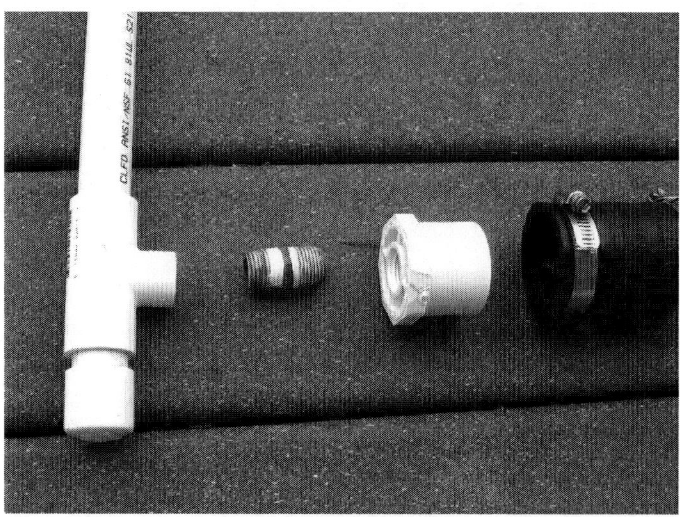

Appendix Figure 5. Reducing bushing and nipple for standpipe.

Appendix Figure 6. Completed drip can.

# Operation of Sprayers for Applying Diluted Liquid Rotenone
# SOP:12.0

**PROCEDURE TITLE:** Operation of Sprayers for Applying Diluted Liquid Rotenone

**APPLICABILITY:** Application of rotenone in waters of the United States

**PURPOSE:** Provide an operating protocol for spray application of prediluted liquid rotenone for fish control

**LOCATION ON LABEL:** "Directions for Use"

**PROCEDURE:**

I. Preparation

Transfer the rotenone concentrate from the original container into the service container or sprayer using the procedure outlined in SOP 10. Sprayers described in this SOP utilize a prediluted (recommended 1 to 2%) solution of liquid rotenone from a 1 to 50-gallon tank. See SOP 8 for spray application of undiluted rotenone concentrate.

II. Operation

A. Generally for stream treatments that do not utilize drip stations, begin spray application at the most downstream portion of the treatment area and work in an upstream direction on both sides of the stream. Generally for stream treatments that utlitize drip stations, follow the flow of rotenone downstream from the drip station on both sides of the stream. To mark locations that have been sprayed, use a dye such as Rhodamine WT in the dilute rotenone mixture or GPS tracking technology. The spray application occurs in concert with the drip stations (if used) dispensing rotenone (see SOP 11).

B. For lake treatments, the spray application of hard to reach areas and weed beds occurs in concert with the general application. Like streams, sprayed locations are marked using a dye such as Rhodamine WT added to the dilute rotenone mixture or GPS tracking technology.

C. Keep track of rotenone amount that is applied to avoid over treatment. Total amount of rotenone applied to the stream or lake from all applications (i.e., drip cans, powder/gelatin/sand mixture, diluted, or undiluted rotenone application) should not exceed the quantity needed for the desired treatment rate.

III. Safety

A. Follow label requirements for PPE for mixers, loaders, applicators and other handlers. All must wear chemically resistant gloves and footwear (or waders), goggles, coveralls, and a dust/mist respirator (see Figure SOP 12.1).

B. The amount of undiluted rotenone required is added to tank directly from a measuring device or from a service container. If added from a service container, make sure that the container is properly labeled (see SOP 4).

C. Make a dilute rotenone solution by partially filling the tank with water, adding the predetermined quantity of undiluted rotenone liquid to the tank, then filling the tank to the desired volume. When filled, it is recommended that the tank is diluted to 1–2% rotenone solution.

D. Following application, the equipment is rinsed by pumping site water for several minutes to rinse liquid mixture from the tanks and hoses. Equipment should be triple-rinsed in site water.

IV. Application

Two commercial types of sprayers are available for applying prediluted liquid rotenone: (1) low-pressure manually-operated backpack and battery-operated sprayers and (2) high-pressure mechanically-operated agricultural sprayers. Prediluted liquid rotenone is sprayed on backwater areas of streams and rivers, seeps, springs and hard to reach shoreline areas and weed beds of ponds and lakes.

A. Mix rotenone from the service container with water at the designated treatment site (see SOP 10).

B. Handle equipment in a manner that minimizes the spillage of liquid rotenone product.

C. Operate the hand-directed nozzle in a manner to minimize small droplets and reduce drift. Using a large-diameter orifice on the spray nozzle, directing the spray downward, and minimizing the distance the spray travels all minimize drift.

V. Equipment

A. Numerous models of manually operated back pack sprayers are commercially available (see Figure SOP 12.2).

B. Several gasoline-powered high-pressure sprayer systems (see Figure SOP 12.3) and lower pressure battery-operated electric spray systems (see Figure SOP 12.4) are commercially available. The larger units are connected to a mixing tank and transported to location with a truck, all-terrain vehicle (ATV), or tractor (see Figure SOP 12.5). Smaller units can be used in a small boat (Figure SOP 12.6).

FIGURE SOP 12.1. Safety gear required for application of diluted rotenone spray using backpack sprayer. Note that applicator is not wearing a dust/mist respirator that is now required.

FIGURE SOP 12.2. Typical manually-operated backpack sprayer.

FIGURE SOP 12.3. Typical gasoline-powered high-pressure sprayer units.

FIGURE SOP 12.4. Typical battery-powered electric spray unit with tank.

FIGURE SOP 12.5. High-pressure sprayer is transported to site with an ATV. Spray is delivered to site with 500 feet of high-pressure hose mounted on a hose reel. Note that applicators are not wearing dust/mist respirators that are now required (Photo credit: California Department of Fish and Game).

FIGURE SOP 12.6. Battery operated sprayer in a small vessel. Note that applicators are not wearing dust/mist respirators that are now required (Photo credit: Washington Department of Fish and Wildlife).

# USE OF ROTENONE POWDER/GELATIN/SAND MIXTURE
# SOP:13.0

**PROCEDURE TITLE:** Use of Rotenone Powder/Gelatin/Sand Mixture

**APPLICABILITY:** Application of rotenone in waters of the United States

**PURPOSE:** Provide guidance on the preparation and use of rotenone powder/gelatin/sand mixture for treating sources of upwelling groundwater in springs, streams and lakes and other areas with limited water circulation (e.g., dense weed beds)

**LOCATION ON LABEL:** "For Use in Springs, Seeps, and Areas with Poor Water Exchange" under the heading "Directions for Use"

**PROCEDURE:**

I. Preparation

Transfer rotenone powder from its original container in an area free of air movement such as an enclosed laboratory or warehouse. Wear all required safety gear on the label for those in contact with the rotenone powder, directly or through drift. If an enclosed area isn't available, or preparing the mixture off-site is logistically not feasible, then a safe area at the treatment site away from other activities should be used. Plan the mixing when air movement is expected to be minimal, set up a work area to contain spills (e.g., blue plastic tarp with a berm), wear appropriate PPE, and prepare the mixture in small quantities (i.e., 5-gallon buckets with lids).

A. Mixture formula is 1 pound (0.454 kg) of powdered rotenone to 1 pound (0.454 kg) of fine-to-medium sand, to 2 ounces (0.0567 kg) of unflavored gelatin (1:1:0.125). Add sufficient water to create a dough-like consistency.

B. Use sand of uniform consistency <1mm diameter and clean (washed or screened to remove organics and silt).

C. Gelatin powder is available in the home canning section of most supermarkets.

D. The finished mixture should form a ball when squeezed tightly, but readily break apart upon contact with the water.

E. The mixture is stored in 5-gallon buckets with air-tight lids to keep the mixture moist until used. Plan to use all the mixture within a few days of mixing because mold will develop in the air tight containers. Storing the buckets in a cool environment will keep the mixture fresh.

II. Operation

A. Prepare small amounts (less than 10 pounds (4.536 kg) total) of the sand mixture by hand in buckets or similar containers. Mixing and dust production are eased by adding no more than 1 pound (0.454 kg) sand and 1 pound (0.454 kg) powdered rotenone to the gelatin at a time, then adding a small amount of water prior to mixing. Mix a small batch while continuing to add water until the proper consistency is obtained. Continue adding the components in small amounts until the desired quantity is mixed.

B. Larger amounts of the mixture can be prepared in a closed container by moistening powdered rotenone in its plastic container, then adding clean sand, and stirring. Water is sprayed on the mixture to moisten the sand and add just enough gelatin to cause the powdered rotenone to adhere to the sand granules. Do not attempt mixing large quantities until proficiency has been developed in mixing smaller amounts, as described in A.

C. One cup (approximately 1 pound (0.454 kg)) of this mixture will treat approximately 0.5 $ft^3/s$ (0.0142 $m^3/s$) of moving water for up to 12 hours at approximately 18 ppb active rotenone (5% a.i. rotenone powder). Use more or less mixture to achieve higher or lower desired rotenone concentrations in treatment area.

III. Safety

A. Wear PPE as required on rotenone powder label for mixers handling powdered rotenone that includes coveralls, chemical resistant gloves and footwear (or waders), protective eyewear, and a NIOSH-approved full-face or hood-style respirator.

B. Wear PPE as required on rotenone powder label for applicators handling the moistened mixture, except a dust/mist respirator can be substituted for a NIOSH-approved full-face or hood-style respirator. Only apply moistened rotenone powder/gelatin/sand mixture, and do not apply dried out mixture.

IV. Application

A. Locate areas of upwelling groundwater and springs by reconnoitering stream sections or using Rhodamine WT dye. Upwelling in lakes may be determined by observing sites where winter ice melts before the rest of the lake and recording locations on charts or GPS.

B. Applicators dispense the mixture by hand over seep and upwelling areas, springs, deep beaver ponds and dense weed beds. In springs with a significant discharge, placing the mixture in a burlap bag weighted down by a rock may keep the rotenone in place during the treatment.

C. Do not exceed the desired concentration of rotenone for the volume or flow of water treated.

D. Empty containers (buckets, backpacks, plastic bags, etc.) used to carry or store the mixture should be triple-rinsed using water in the treatment area to remove residue.

## V. Additional Information

Spateholts, R. L., and L. E. Lentsch. 2001. Utah's rotenone sandmix: a formulation to maintain fish toxicity in seeps or springs. Pages 107–118 *in* R. L. Cailteux, L. DeMong, B. J. Finlayson, W. Horton, W. McClay, R. A. Schnick, and C. Thompson, editors. Rotenone in fisheries: are the rewards worth the risks? American Fisheries Society, Trends in Fisheries Science and Management 1, Bethesda, Maryland.

# USE OF *IN-SITU* BIOASSAYS TO MONITOR EFFICACY
# SOP:14.0

**PROCEDURE TITLE:** Use of *In-Situ* Bioassays to Monitor Efficacy

**APPLICABILITY:** Application of rotenone in waters of the United States

**PURPOSE:** Provide protocol and rationale for the use of *in-situ* bioassays to monitor the efficacy of rotenone treatments and potassium permanganate ($KMnO_4$) deactivation.

**LOCATION ON LABEL:** "Determining Treatment Rate," "Drinking Water Monitoring," and "Placarding of Treatment Areas" under the heading "Directions for Use"

**PROCEDURE:**

I. Considerations

A. Both $KMnO_4$ and rotenone are toxic to fish, and both act more quickly at higher temperatures and more slowly at lower temperatures. The basis for using bioassays to monitor the toxicity of $KMnO_4$ or rotenone is the concept of the dose-response relationship. The time-dependence of the toxicity of $KMnO_4$ to rainbow trout is shown by the Marking and Bills (1975) study in Figure SOP 14.1. As the dose (or concentration) of rotenone or $KMnO_4$ is increased, the time needed to cause an effect is decreased. Loeb and Engstrom-Heg (1971) demonstrated this concept for brown trout and rotenone in Figure SOP 14.2. The response is shortened with increasing dose and temperature.

B. Bioassays to monitor efficacy of a treatment are typically done *in-situ*, by placing fish in some enclosure (cage, bucket or net bag) within the treated waters. Cages used for this purpose typically use ¼-inch nylon mesh netting tied to a hard metal or plastic frame. Buckets are typically plastic, hold 3 to 5 gallons, have lids and ¼-inch holes drilled in the walls to allow for rapid water exchange. Net bags are typically ¼-inch mesh netting with a drawstring at one end to prevent escape of the fish. These net bags (sometimes sold as bait bags) can come with or without metal hoops to keep the void of the net open to allow fish movement. Ideally, the fish should be of a range of sizes representative of the fish species being killed during the treatment, and generally only several fish are needed in each container.

C. Bioassays require a fish species of similar or lesser sensitivity than the target species. Because it is always possible that fish might escape their holding chamber, it is paramount to use holding devices in a state of good repair, provide security, and account for all fish that are used. It is preferable not to use the species that the treatment is trying to eradicate, if possible. Ideally, the species to use is that being restocked in the water body after the treatment, assuming there is similar or lesser sensitivity to rotenone.

Another option would be to do the bioassays *ex-situ*, on-site but away from the stream margin, by placing fish of a closely-related species with similar tolerances to rotenone in a bucket with treated stream water. Replenishment of water in this bucket should be done at least hourly and artificial aeration can be provided if necessary although this may speed the breakdown of rotenone. *Ex-situ* bioassays may not provide an accurate assessment of the treatment due to different exposure conditions than that in the stream environment. One other option would be to use a sterile (triploid) congener, where escapees would not be able to contribute genetic material to the new species being stocked into the waterbody. Sterile fish might also be the choice in situations where the intention is to manage the treated waters in a fishless state.

## II. Application

### A. Rotenone Treatments

#### 1. Flowing Waters

Flowing waters are typically treated using drip stations, and a caged-fish bioassay should be conducted as far downstream as the toxic effects are expected to occur. When numerous drip stations are operated at regular intervals along the stream (1 to 2 hours travel-time apart) the cages should be placed immediately upstream of each drip station to evaluate the efficacy of the treatment from the upstream station. The death of caged fish within the duration of the treatment (typically 4–8 hours) is used to verify that the rotenone treatment is adequate and that enough rotenone carried the full distance to the next lower drip station to assure a complete kill. On large projects where there may be many drip stations, backpack sprayers or rotenone/sand/gelatin mixture are used concurrently, it may be a matter of concern whether rotenone concentrations are remaining stable, increasing or decreasing through the treatment area. If this occurs, it will be most evident at the lower end of the treatment area, and the response time of caged fish placed just above the deactivation station will allow the applicator to track any changes in the concentration of rotenone.

#### 2. Standing Water Treatments

Evaluate efficacy at all depths throughout the water body. Suspend fish in cages or net bags at the surface, immediately off the bottom, and some mid-depth location, preferably right below the thermocline if one exists. In open water situations, wait one day before deploying the cages to allow the chemical to mix first. Alternatively, place fish in cages in the water body the day before treatment, but recognize that variable mixing throughout the water column affects response time for the fish. Check fish after 2 days of exposure. If the treatment is done in winter below the ice, wait a week to allow mixing before deploying the cages.

B. Deactivation Operations

Use bioassays to determine the efficacy of $KMnO_4$ in deactivating rotenone and when to terminate the deactivation. Place bioassay fish immediately upstream of $KMnO_4$ introduction site and at 15-minute and 30-minute travel-time distances further downstream.

1. Upstream of Deactivation

Place fish upstream to indicate the need for beginning of deactivation operation and after the treatment to indicate the cessation of the deactivation operation. Place fish in the stream immediately above the site of the $KMnO_4$ application. Deactivation operation is started when fish show signs of stress at the beginning of the treatment. As an alternative, begin deactivation when the rotenone treatment begins. Continue the deactivation operation until the application of rotenone has stopped and after the hypothetical clearing time for rotenone from the stream. The clearing time is the travel-time from the most upstream rotenone drip station or application to the deactivation station. Reintroduce fish into the cages at that time to establish they are able to survive for 4 hours without any signs of stress.

2. Downstream of Deactivation

Place caged fish in, and downstream of, the mixing (contact) zone to show the efficacy of the $KMnO_4$ deactivation. Caged fish should all survive below the 30-minute contact zone and may survive at the 15-minute mark if deactivation is successful. For long deactivation operations, caged fish should be replaced daily from the stress of confinement and $KMnO_4$. If any of the fish die at the 30-minute mark, then there may be an imbalance between the concentration of $KMnO_4$ and rotenone, and the application rate of $KMnO_4$ may have to be adjusted (see SOP 7).

III. Additional Information

Loeb, H. A., and R. Engstrom-Heg. 1971. Estimation of rotenone concentration by bioassay. New York Fish and Game Journal 18(2):129–134.

Marking, L., and T. Bills. 1975. Toxicity of potassium permanganate to fish and its effectiveness in detoxifying antimycin. Transactions of the American Fisheries Society 104:579–583.

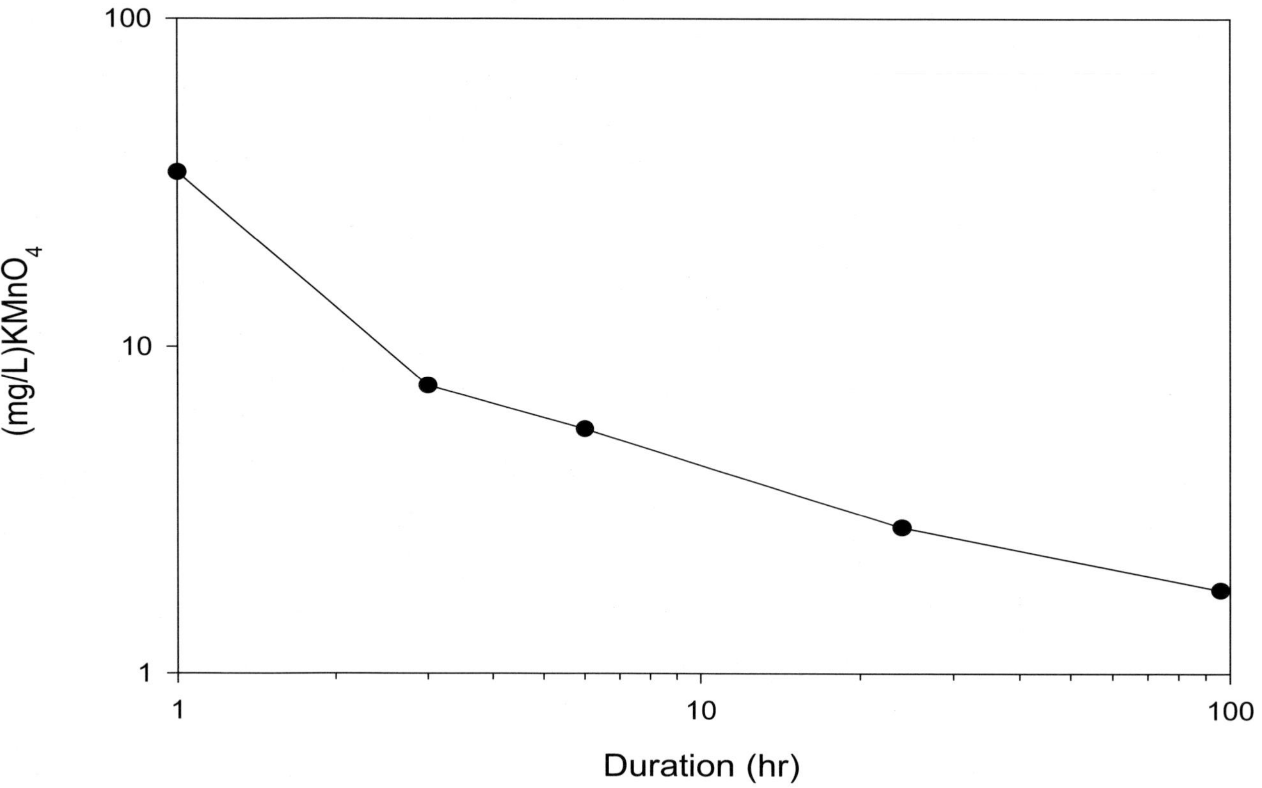


FIGURE SOP 14.1. Acute toxicity (LC50 values) of potassium permanganate to rainbow trout *Oncorhynchus mykiss* at temperature of 12°C and pH of 7.5 (from Marking and Bills 1975).

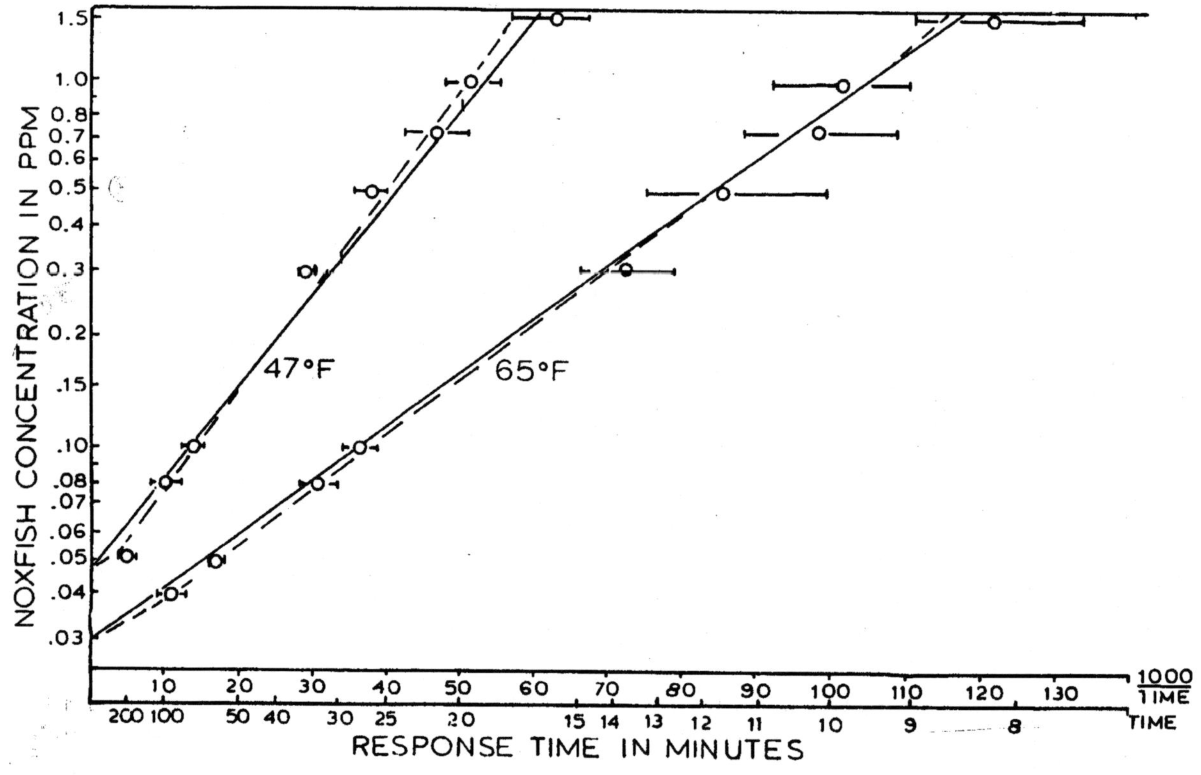


FIGURE SOP 14.2. Response time for brown trout *Salmo trutta* to Noxfish (from Loeb and Engstrom-Heg 1971).

# Collection and Disposal of Dead Fish
# SOP:15.0

**PROCEDURE TITLE:** Collection and Disposal of Dead Fish

**APPLICABILITY:** Application of rotenone in waters of the United States

**PURPOSE:** Provide guidance for the collection and disposal of dead fish

**LOCATION ON LABEL:** "General Application Precautions and Restrictions" under the heading "Directions for Use"

**PROCEDURE:**

I. Considerations

Good management of the process for collecting and disposing of dead fish reduces public relations problems and avoids the appearance of a fish kill beyond the intended treatment area. State agencies do not collect and remove dead fish for disposal under normal circumstances. The policy of the California Department of Fish and Game (1994) is to remove fish from the treatment area when dead fish may become a public nuisance or when a request is made by a public agency. The Michigan Department of Natural Resources (1993) and the Idaho Department of Fish and Game (Horton 1997) normally do not recover dead fish. The Washington Department of Fish and Wildlife normally leaves fish carcasses in the water to provide nutrients for growth of phytoplankton and zooplankton but in response to local concerns removes dead fish that have washed onto the shore of lakeside residences (Temple and Anderson 2008). All contacted state agencies reported that dead fish are recovered to avoid serious public controversy in sensitive situations. Cooperate with other agencies and entities when removing fish. Any disposal of fish or other project materials on federal land must be coordinated in advance with the federal land management agency for necessary approvals.

Plan to have adequate resources available for the collection of dead fish. Do not offer or provide dead fish for human consumption because no tolerance for rotenone in fish flesh for human consumption has been established. Additionally, there are other public health issues (e.g., flies and *Salmonella*) associated with decaying flesh.

II. Collection

Collect fish during and after standing water treatments by crews in boats and/or walking the shore with dipnets. Because most fish carcasses sink the collection of dead fish is limited. Bradbury (1986) reported that only about 30% of the dead fish could be recovered from treated lakes in Washington, depending on fish species and size, water depth and temperature, and presence of aquatic vegetation. Pick up fish for two to five days depending on water temperature.

Collect fish from stream treatments using block nets. Monitor the nets and remove fish regularly. Too many fish in the nets may cause collapse of the block nets and loss of the fish downstream.

## III. Disposal

Transport fish to a prearranged disposal site. In wilderness, fish can be buried on-site, away from the stream. Fish will likely be unearthed by bears and other carrion feeders, especially if the grave site is large or shallow. Before digging a hole for disposal, check with the responsible land use agency to determine if an archeological clearance is required. Any ground disturbance on National Forest lands requires approval from the local District Ranger, an archeological clearance and a Categorical Exclusion (NEPA) completed as part of the approval process for the treatment; factor in lead time to accomplish this approval.

Off-site, fish will usually be accepted at sanitary landfills. There may be issues with transporting a large number of dead fish including smell and animal waste on roads. Prior planning and contacting landfill operators before the treatment is prudent.

## IV. Additional Information

Bradbury, A. 1986. Rotenone and trout stocking: a literature review with special reference to Washington Department of Game's lake rehabilitation program. Washington Department of Game, Fisheries Management Report 86-2.

California Department of Fish and Game. 1994. Rotenone Use for Fisheries Management: July 1994. Final Programmatic Environmental Impact Report (Subsequent). State of California Department of Fish and Game.

Horton, W. D. 1997. Federal Aid in Sport Restoration fishery management program, lake renovation manual. Idaho Department of Fish and Game (IDFG 97-9), Boise.

Michigan Department of Natural Resources. 1993. Policy and procedures of the use of piscicides and other compounds by the Fisheries Division in ponds, lakes, and streams. Michigan Department of Natural Resources, Fisheries Division, Lansing.

Temple, R., and J. Anderson. 2008. Draft programmatic environmental assessment for Washington Department of Fish and Wildlife statewide lake and stream rehabilitation program as funded by the USFWS Wildlife and Sportfish Restoration Program. U.S. Fish and Wildlife Service, Portland, Oregon.

# Monitoring Requirements for Aquaculture and Drinking Water
# SOP:16.0

**PROCEDURE TITLE:** Monitoring Requirements for Aquaculture and Drinking Water

**APPLICABILITY:** Application of rotenone in waters of the United States

**PURPOSE:**
1. Mitigate exposure resulting from consuming treated fish or from drinking treated water
2. Provide guidance on the monitoring requirements and analysis of water samples for rotenone concentration

**LOCATION ON LABEL:** "Monitoring and Notification Requirements for Water" under the heading "Directions for Use"

**PROCEDURE:**

Waters treated with rotenone and used for food production of aquatic species (aquaculture) require the collection and analysis of water samples for verifying that rotenone is no longer present (<2 ppb (<0.002 ppm) rotenone) prior to restocking of the treated water with fish. Waters treated with rotenone and used for drinking or with hydrologic connections to wells, when application rate is ≥40 ppb (≥0.04 ppm) rotenone, require the user to be advised against the consumption of water until: (1) active rotenone is <0.04 ppm as determined by analytical chemistry, or (2) fish of the Salmonidae or Centrarchidae families can survive for 24 hours, or (3) dilution with untreated water yields a calculation that active rotenone is <0.04 ppm, or (4) distance or travel-time from the application site are known to produce an active rotenone concentration that is <0.04 ppm.

I. Monitoring Requirements for Aquaculture

The Certified Applicator or designee under his/her direct supervision must prohibit restocking of fish until monitoring samples confirm rotenone concentrations are below the level of detection (<2 ppb) for 3 consecutive samples taken no less than 4 hours apart.

A. Collection of Water Samples for Rotenone Analysis

1. Collect water samples at depth using a Kemmerer bottle or by submerging the sample bottle directly a few inches below the surface. Before taking a sample, the Kemmerer bottle is triple-rinsed with water from sampling depth.

2. Collect samples for analysis of rotenone in chemically clean, 250-ml amber glass bottles with Teflon-lined caps. Exclude air space in the sample bottles and caps.

3. Store samples on ice or cooled to a temperature of 4°C in the dark while in transit to and storage at the laboratory.

4. Laboratory-specific chain-of-custody forms accompany samples documenting the sequence of transfer from collection to chemical analysis.

5. Analyze samples within the acceptable holding time of 6 days.

B. Analysis of Water Samples for Rotenone Concentration

1. Analyze samples for rotenone by liquid chromatography (LC) as described by Dawson et al. (1983) or by direction injection liquid chromatography/ mass spectrometry (LC/MS) as described by McMillin and Finlayson (2008).

2. Both procedures yield a minimum detection limit (MDL) of 1 ppb and reporting limit (RL) of 2 ppb.

II. Requirements for Drinking Water Treated at ≥40 ppb Rotenone Requires Notification Until:

A. Collection and Analysis of Water Samples for Rotenone Analysis

1. See procedure for aquaculture above.

2. A water sample result showing rotenone concentration <40 ppb does not require the advisement against consumption.

**OR**

B. Bioassay Water with Fish of Salmonidae or Centrarchidae Families

1. See SOP 14 for detailed instructions on conducting bioassays.

2. Salmonid or Centrarchid fishes surviving for 24 hours do not require the advisement against consumption because these species cannot survive concentrations ≥40 ppb rotenone.

**OR**

C. Dilution with Untreated Water

1. Calculate a dilution ratio to derive fraction of treated water diluted with untreated water and multiply that fraction by the rotenone treatment dose to determine expected concentration of rotenone.

2. For example: a stream discharge of 5 $ft^3/s$ treated at 60 ppb rotenone flows into an untreated stream discharge of 15 $ft^3/s$. The fraction of treated to untreated water is 5 $ft^3/s \div (5\ ft^3/s + 15\ ft^3/s)$ or 0.25. The combined discharge of the two streams would have an expected concentration of 15 ppb

(60 x 0.25) rotenone following dilution with untreated stream water. A demonstrated concentration <40 ppb rotenone does not require advisement against consumption.

**OR**

D. Distance or Travel-Time from Application Site

1. Generally, rotenone neutralizes over distance and time so that significant residues would not be expected more than several miles or several hours travel time from the application site. Higher temperatures, solar radiation, turbulence, turbidity, and pH will increase the breakdown and dissipation of rotenone over distance and time.

2. This would be expected to vary among water bodies and cannot be predicted with any confidence. To know with certainty, conduct a pilot test in the water body. Apply rotenone at the anticipated rate and sample rotenone in water below an application site and at multiple spots downstream. The sample site where the rotenone concentration drops below 40 ppb represents the distance or travel time needed in order to avoid the notification requirement. This distance or travel time can then be applied to any treatments in that water body at different times or locations.

III. Additional Information

Dawson, V., P. Harmon, D. Schultz and J. Allen. 1983. Rapid method for measuring rotenone in water at piscicidal concentrations. Transactions of American Fisheries Society 112:725–727.

McMillin, S., and B. Finlayson. 2008. Chemical residues and sediment following rotenone application to Lake Davis, California, 2007. California Department of Fish and Game Pesticide Investigations Unit, Office of Spill Prevention and Response Administrative Report 08-01.

# 4 References


Allendorf, F. 1991. Ecological and genetic effects of fish introductions; synthesis and recommendations. Canadian Journal of Fisheries and Aquatic Sciences 48 (Supplement 1):178–181.

Ball, R. 1948. A summary of experiments in Michigan lakes on the elimination of fish populations with rotenone, 1934–1942. Transactions of American Fisheries Society 75:139–146.

Betarbet, R., T. Sherer, G. MacKenzie, M. Garcia-Osuna, A. Panov, and J. Greenamyre. 2000. Chronic systemic pesticide exposure reproduces features of Parkinson's disease. Nature Neuroscience 3(12):1301–1306.

Bills, T., and L. Marking. 1986. Rotenone—freshwater $LC_{50}$—rainbow trout and bluegills. U.S. Fish and Wildlife Service, National Fish Research Laboratory, La Crosse, Wisconsin. Project TOX 83–626.01B.

Bills, T. D., J. J. Rach, and L. L. Marking. 1986. Toxicity of rotenone to developing rainbow trout. U.S. Fish and Wildlife Service, National Fish Research Laboratory, La Crosse, Wisconsin. June 1986.

Biotech Research Laboratories, Inc. 1981. Analytical studies for the detection of chromosomal aberrations in fruit flies, rats, mice and horse bean. Biotech Research Laboratories, Inc. Rockville, Maryland.

CDFG (California Department of Fish and Game). 1991. Northern pike eradication project—draft subsequent environmental impact report (SCH 92073015). Environmental Services Division, California Department of Fish and Game, Sacramento.

CDFG. 1994. Rotenone use for fisheries management—final programmatic environmental impact report (SCH 92073015). CDFG, Environmental Services Division, Sacramento.

CDFG. 1997. Lake Davis Northern pike eradication project—January 1997 (SCH 95022026). California Department of Fish and Game, Region 2 Headquarters, Rancho Cordova.

CDFG. 2007. Lake Davis Pike Eradication Project—Final EIR/EIS. January, 2007 (SCH 2005-09-2070). California Department of Fish and Game, Sacramento.

Cheng, H., I. Yammamoto, and J. Casida. 1972. Rotenone photodecomposition. Journal of Agricultural Food and Chemistry 20:850–856.

Cumming, K. B. 1975. History of fish toxicants in the United States. Pages 5–21 *in* P. H. Eschmeyer, editor. Rehabilitation of fish populations with toxicants: a symposium. American Fisheries Society, North Central Division, Special Publication 4, Bethesda, Maryland.

Dawson, V. 1986. Adsorption-desorption of [6a-$^{14}$C] by bottom sediments. U.S. Fish and Wildlife Service, National Fishery Research Laboratory, La Crosse, Wisconsin. Report ROT-84-988.02.

Draper, W. 2002. Near UV quantum yields for rotenone and piperonyl butoxide. Analyst 127:1370–1374.

Eiseman, J. L. 1984. General metabolism study for safety evaluation of rotenone using rats. Project No. 419–137. Hazleton Laboratories America Inc., Vienna, Virginia.

Ellis, H. V., S. Unwin, J. Cox, I. S. Elwood, E. A. Castillo, E. R. Ellis, and J. Carter. 1980. Subchronic oral dosing study for safety evaluation of rotenone using dogs. Final report. Midwest Research Institute, Kansas City, Missouri.

EPA (U.S. Environmental Protection Agency). 2007. Reregistration Eligibility Decision for Rotenone EPA 738-R-07-005. U.S. EPA, Prevention, Pesticides, and Toxic Substances, Special Review and Reregistration Division, March 2007.

EPIWIN (v.3.12) 2004. U.S. EPA OPPT and Syracuse Research Corporation. Available: http://www.epa.gov/opptintr/exposure/docs/episuited1.htm.

Fang, N., and J. Casida. 1999. Cube resin insecticide: identification and biological activity of 29 rotenoid constituents. Journal of Agricultural Chemistry 47(5):2130–2136.

Fang, N., J. Rowlands, and J. Casida. 1997. Anomalous structure-activity relationships of 13-*homo*-oxatorenoids and 13-*homo*-oxadehydrorotenoids. Chemical Research and Toxicology 10(8):853–858.

Finlayson, B., R. Schnick, R. Cailteux, L. DeMong, W. Horton, W. McClay, C. Thompson, and G. Tichacek. 2000. Rotenone use in fisheries management: administrative and technical guidelines manual. American Fisheries Society, Bethesda, Maryland.

Finlayson, B., W. Somer, D. Duffield, D. Propst, C. Mellison, T. Pettengill, H. Sexauer, T. Nesler, S. Gurtin, J. Elliot,

F. Partridge, and D. Skaar. 2005. Native inland trout restoration on National Forests in the Western United States: Time for Improvement. Fisheries 30(5):10–19.

Finlayson, B., J. Trumbo, and S. Siepmann. 2001. Chemical residues in surface and ground waters following rotenone application to California lakes and streams. Pages 37–53 *in* R. C. Cailteux, L. DeMong, B. J. Finlayson, W. Horton, W. McClay, R. A. Schnick, and C. Thompson, editors. Rotenone in fisheries: are the rewards worth the risks? American Fisheries Society, Trends in Fisheries Science and Management 1, Bethesda, Maryland.

Finlayson, B. J., R. A. Schnick, R. L. Cailteux, L. DeMong, W. D. Horton, W. McClay, and C. W. Thompson. 2002. Assessment of Antimycin A use in fisheries and its potential for reregistration. Fisheries 27(6):10–18.

Freudenthal, R. I. 1981. Carcinogenic potential of rotenone: subchronic oral and peritoneal administration to rats and chronic dietary administration to Syrian golden hamsters. Health Effects Research Laboratory, Office of Research and Development, US Environmental Protection Agency, North Carolina.

Gabriel, D. 1996. Acute dermal toxicity, single level—rabbits. Project No. 95-8249A. Bioresearch Incorporated, Philadelphia, Pennsylvania.

Gingerich, W., and H. J. Rach. 1985. Uptake, biotransformation, and elimination of rotenone by bluegills (*Lepomis macrochirus*). Aquatic Toxicology, 6:179–196.

Greenman, D. L., W. Allaben, G. Burger, and R. Kodell. 1993. Bioassay for carcinogenicity of rotenone in female Wistar rats. Fundamental and Applied Toxicology 20:383–390.

Haley, T. 1978. A review of the literature of rotenone. Journal of Environmental Pathology and Toxicology 1:315–337.

Hansch, C., A. Leo, and D. Hoekmann. 1995. Exploring QSAR. Hydrophobic, Electronic and Steric Constants. *In* S. R. Heller, editor. American Chemical Society Professional Reference Book. Washington, D.C.

Haworth, S. R. 1978. Salmonella/mammalian—microsome plate incorporation mutagenesis assay (rotenone). Study No. 019-563-165-1. EG&G Mason Research Institute, Rockville, Maryland.

Hobert, M. 1995. Acute inhalation toxicity study in rats with AEH #899. Laboratory Study No. 1954–95. Stillmeadow Inc., Sugar Land, Texas.

Höglinger, G. U., W. H. Oertel and E. C. Hirsch. 2006. The rotenone model of Parkinsonism—the five years inspection. Journal Neural Transmission Supplement 70:269–72.

Hollingworth, R. M. 2001. Inhibitors and uncouplers of mitochondrial oxidative phosphorylation. Pages 1169–1263 *in* R. Krieger, editor. Handbook of Pesticide Toxicology, 2nd Edition, Academic Press, New York.

Huntingdon Life Sciences, Ltd. 2007. Rotenone, chemical-physical properties. Cambridgeshire, PE28, 4HS, U.K.

Johnsen, B. O., Å. Brabrand, P. A. Jansen, H. -C. Teien, and G. Bremset. 2008. Evaluering av bekjempelsesmetoder for *Gyrodactylus salaris*. Rapport fra ekspertguppe. Uttredning for DN 200807.

Kaeding, L. R., G. D. Boltz, and D. G. Carty. 1996. Lake trout discovered in Yellowstone Lake threaten native cutthroat trout. Fisheries 21:16–20.

Krueger, C. C., and B. May. 1991. Ecological and genetic effects of salmonid introductions in North America. Canadian Journal of Fisheries and Aquatic Sciences 48:66–77.

Kuhn, J. 1995. Dermal sensitization study in guinea pigs with SHE #899. Stillmeadow, Inc., Sugar Land, Texas. AgrEvo Environmental Health Study T-95-101.

Lemly, A. D. 1985. Suppression of native fish populations by green sunfish in first-order streams of Piedmont North Carolina. Transactions of the American Fisheries Society 114:705–712.

Lennon, R. E., J. B. Hunn, R. A. Schnick, and R. M. Burress. 1970. Reclamation of ponds, lakes, and streams with fish toxicants: a review. Food and Agriculture Organization of the United Nations, Fisheries Technical Paper 100.

Lowe, C. 2006a. CFT Legumine, acute oral toxicity up and down procedure in rats. Eurofins Product Safety Laboratories, Dayton, New Jersey.

Lowe, C. 2006b. CFT Legumine, acute inhalation study in rats—limit test. Eurofins Product Safety Laboratories, Dayton, New Jersey.

MacKenzie, K. M. 1981. Teratology study with rotenone in mice. Raltech Scientific Services, Madison, Wisconsin.

MacKenzie, K. M. 1982. Teratology study with rotenone in rats. Hazleton Raltech Inc., Madison, Wisconsin.

MacKenzie, K. M. 1983. Reproduction study for safety evaluation of rotenone using rats. Final report. Hazleton Raltech, Inc. Madison, Wisconsin.

Meronek, T. G., P. M. Bouchard, E. R. Buckner, T. M. Burri, K. K. Demmerly, D. C. Hatleli, R. A. Klumb, S. H. Schmidt, and D. W. Coble. 1996. A review of fish control projects. North American Journal of Fisheries Management 16:63–74.

McClay, W. 2000. Rotenone use in North America (1988–1997). Fisheries 20(5):15–21.

McClay, W. 2005. Rotenone use in North America (1988–2002). Fisheries 30(4):29–31.

McMillin, S., and B. Finlayson. 2008. Chemical residues in water and sediment following rotenone application to Lake Davis, California, 2007. California Department of Fish and Game, Office of Spill Prevention and Response Administrative Report No. 08-01. Sacramento.

MDNR (Michigan Department of Natural Resources). 1990. An assessment of human health and environmental effects of use of rotenone in Michigan's fisheries management programs. MDNR, Fisheries Division, Lansing.

Moore, G. 1995a. Primary skin irritation—rabbits: AEH 897 (crystalline rotenone). RUC Study T95-104, Project Number 95-8249A. Bioresearch Incorporated, Philadelphia, Pennsylvania.

Moore, G. 1995b. Primary eye irritation, 6 washed and 3 washed—rabbits—AEH 897. Bioresearch Incorporated, Philadelphia, Pennsylvania. Bioresearch Project Number 95-8249A.

National Toxicology Program 1984. Mouse Lymphoma Protocol (Rotenone). Laboratory Project No. 28037. National Toxicology Program, Research Triangle Park, North Carolina.

National Toxicology Program. 1986. Toxicology and carcinogenesis studies of rotenone (CAS No. 83-79-4) in F344/N rats and $B6C3F_1$ mice (feed studies). NTP Technical Report Series No. 320. Triangle Park, North Carolina.

Rach, J. J., T. D. Bills, and L. L. Marking. 1988. Acute and chronic toxicity of rotenone to Daphnia magna. U.S. Fish and Wildlife Service, Investigations in Fish Control 92:1–5.

Reigart, J. and J. Roberts. 1999. Recognition and management of pesticide poisoning. Fifth Edition, EPA-735-R-98-003, Washington, D.C.

Rojo, A. I., C. Cavada, M. Rosa de Sagarra, and A. Cuadrado. 2007. Chronic inhalation of rotenone or paraquat does not induce Parkinson's disease symptoms in mice or rats. Experimental Neurology 208(1):120–126.

Ross, S. T. 1991. Mechanisms structuring stream fish assemblages: are there lessons from introduced species? Environmental Biology of Fishes 30:359–368.

Schnick, R. A. 1974. A review of the literature on the use of rotenone in fisheries. U.S. Department of Interior, U.S. Fish and Wildlife Service, Fish Control Laboratory, National Technical Information Service PB-235 454, La Crosse, Wisconsin.

Sherer, T., R. Betarbet, C. Testa, B. Seo, J. Richardson, J. Kim, G. Miller, T. Yagi, A. Yagi and J. Greenamyre. 2003. Mechanisms of toxicity in rotenone models of Parkinson's disease. Journal of Neuroscience 23(34):10756–10764.

Solman, V. 1950. History and use of fish poisons in the United States. Canadian Fish Culturist 8:3–16.

Spare, W. 1982. Aqueous photodegradation of $^{14}$C-rotenone. Biospherics Inc., Rockville, Maryland, submitted by Litton Bionetics, Inc., Kensington, Maryland. LBI Project No. 22147-01.

Steele, J. 1982. Effects of rotenone on microflora of soil, sediment, and water. Hazelton Raltech, Inc. Madison, Wisconsin. Study No 81503.

Stevenson, J. 1978. The acute toxicity of unformulated pesticides to worker honey bees. Plant Pathology 27(1):38–40.

Swan, G. 2007. CFT Legumine 5% in vitro dermal penetration using human skin. Huntingdon Life Sciences, Cambridgeshire, PE28, 4HS, U.K.

Thomas, R. 1983. Hydrolysis of [6-$^{14}$C]-rotenone. Borriston Laboratories, Inc. Borriston Project No. 0301A.

Tisdale, M. 1985. Chronic toxicity study of rotenone in rats. Hazelton Laboratories American, Madison, Wisconsin. Study No. 6115–100

Tomlin, C. 1994. The pesticide manual. British Crop Protection Council, London, England.

Tucker , R. 1968. Rotenone: oral toxicity to mallard ducks and ring-necked pheasants. Unpublished study prepared by U.S. Fish and Wildlife Service, Wildlife Research Center, Denver, Colorado.

WDFW (Washington Department of Fish and Wildlife). 1992. Environmental impact statement for lake and stream rehabilitations, 1992–1993—final supplemental report. WDFW, Habitat and Fisheries Management Divisions, Report 92–14, Olympia.

Westervelt, J. 2007. Lake Davis Pike Eradication Project personal air monitoring report. California Department of Fish and Game, Office of Spill Prevention and Response, Sacramento.

Williams, T. 2004. Environmentalists vs. Native Trout. Fly Rod and Reel. 2004 (April):18–26.

Zimmerman, M. P., and D. L. Ward. 1999. Index of predation on juvenile salmonids by northern pikeminnow in the lower Columbia River basin, 1994–1996. Transactions of the American Fisheries Society 128:995–1007.